678.5.033

R.O. Westoctt

T.I.B./Library

_____

Borrower   Return by
(126)
Mr N Evans 80.6.90
            20.12.96

                20.6.91
                2.12.91
Mr N Evans 6.6.92
Mr N Evans 26/12/92

DEVELOPMENTS IN COMPOSITE MATERIALS—1

DEVELOPMENTS IN COMPOSITE MATERIALS—1

Edited by

G. S. HOLISTER

*Faculty of Technology, The Open University,
Walton Hall, Milton Keynes, UK*

APPLIED SCIENCE PUBLISHERS LTD
LONDON

APPLIED SCIENCE PUBLISHERS LTD
RIPPLE ROAD, BARKING, ESSEX, ENGLAND

ISBN: 0 85334 740 9

WITH 23 TABLES AND 68 ILLUSTRATIONS

© APPLIED SCIENCE PUBLISHERS LTD 1977

Printed in Great Britain by Galliard (Printers) Limited, Great Yarmouth

PREFACE

At their inception fibre-reinforced composites were hailed by the press and the popular scientific journals as a major technological breakthrough which was to revolutionise within a few years our use of materials in structures and machines. Extravagant claims were made for their outstanding physical properties and low costs. It was predicted that the time was imminent when material properties would be as much subject to design specification and modification as the product of which they were a part.

It was inevitable that with the first major technological setback—their demonstrated unsuitability for use in the latest generation of turbo-jet engines—a reaction should set in, and it is arguable that fibre-reinforced composites are as much undervalued now as they were over-sold at their inception. Over the past few years technological progress has continued quietly and unspectacularly to improve the physical properties of such composites and, equally important, also to improve our understanding of how they may be used to greatest effect. At the core of this progress has been extensive research into the behaviour of composite materials under realistic environmental conditions (corrosive environments, vibrational stress, extreme conditions of loading), and the nature of the complex stress fields that occur in such materials due to their configurations, loading conditions and internal physical structure.

The papers in this book are representative of the progress that is being made in these areas, and indicate the degree to which our understanding of fibre-reinforced composites is gradually progressing to a stage at which, perhaps, composite materials will eventually fulfil the initial claims that were made for them.

CONTENTS

Preface v

List of Contributors ix

1. Mechanisms of Corrosion in Carbon Fibre-Reinforced Phenolic Resins 1
   G. L. HART and G. PRITCHARD

2. Rupture During Bending of Composites Reinforced with Discontinuous and Unidirectional Metal Fibres: Study Based on a Recorded Test 17
   A. CUPCIC and J. M. BERTHELOT

3. Vibrations of Antisymmetrically Laminated Cylindrical Shells 37
   Y. V. K. SADASIVA RAO and P. C. RAJU

4. Effect of Shear Deformation and Anisotropy on the Non-Linear Response of Composite Plates 55
   A. K. NOOR and S. J. HARTLEY

5. Experimental–Numerical Hybrid Technique for Stress Analysis of Orthotropic Composites 67
   K. CHANDRASHEKHARA and K. ABRAHAM JACOB

6. Property Optimisation Analysis for Multicomponent (Hybrid) Composites 85
   R. L. MCCULLOUGH and J. M. PETERSON

7. Constitutive Relationships for Heterogeneous Materials . 119
   C.-T. D. Wu and R. L. McCullough

8. Stress Concentrations Around Circular Holes in a Filament
   Stiffened Sheet: Some Experimental Results . . . 189
   D. J. Malcolm

9. Stresses in Three-Dimensional Composites with Limiting Shear
   Properties 197
   P. S. Theocaris and S. A. Paipetis

10. A Plane Strain Solution of an Elastic Cylindrical Inclusion in an
    Elasto-Plastic Matrix Under Uniform Tension . . . 209
    D. K. Brown

11. Formation of Permanently Curved Boron Filaments . . 227
    M. Schoppee and J. Skelton

Index 241

LIST OF CONTRIBUTORS

K. Abraham Jacob

Department of Civil Engineering, Indian Institute of Science, Bangalore 560012, India.

J. M. Berthelot

Laboratoire de Physique des Matériaux, Centre Universitaire, Route de Laval, 72017 Le Mans Cedex, France.

D. K. Brown

Department of Mechanical Engineering, The Engineering Laboratories, The University, Glasgow G12 8QQ, UK.

K. Chandrashekhara

Department of Civil Engineering, Indian Institute of Science, Bangalore 560012, India.

A. Cupcic

Laboratoire de Physique des Matériaux, Centre Universitaire, Route de Laval, 72017 Le Mans Cedex, France.

G. L. Hart

Unilever Research Laboratories, 455 London Road, Isleworth, Middlesex, UK.

ix

J. Hartley

School of Engineering and Applied Science, George Washington University, National Aeronautics and Space Administration, Langley Research Center, Mail Stop 169, Hampton, Virginia 23665, USA.

D. J. Malcolm

Department of Mechanical Engineering, The University of Calgary, 2920 24 Ave N.W., Calgary, TN1 1N4, Canada.

R. L. McCullough

Department of Chemical Engineering, University of Delaware, Newark, Delaware 19711, USA.

A. K. Noor

School of Engineering and Applied Science, George Washington University, National Aeronautics and Space Administration, Langley Research Center, Mail Stop 169, Hampton, Virginia 23665, USA.

S. A. Paipetis

Department of Mechanics, National Technical University of Athens, Athens 625, Greece.

J. M. Peterson

Boeing Commercial Airplane Company, Seattle, Washington, USA.

G. Pritchard

School of Chemical and Physical Sciences, Kingston Polytechnic, Penrhyn Road, Kingston upon Thames KT1 2EE, Surrey, UK.

P. C. Raju

Structural Engineering Division, Vikram Sarabhai Space Centre, Trivandrum-695022, India.

Y. V. K. Sadasiva Rao

Structural Engineering Division, Vikram Sarabhai Space Centre, Trivandrum-695022, India.

M. M. Schoppee

F.R.L., Route 128 *at U.S.* 1, *Dedham, Massachusetts* 02026, *USA.*

J. Skelton

F.R.L., Route 128 *at U.S.* 1, *Dedham, Massachusetts* 02026, *USA.*

P. S. Theocaris

Department of Mechanics, National Technical University of Athens, Athens 625, *Greece.*

C.-T. D. Wu

Mobil Research and Development Corporation, Research Department, Paulsboro, New Jersey 08066, *USA.*

Chapter 1

MECHANISMS OF CORROSION IN CARBON FIBRE-REINFORCED PHENOLIC RESINS

G. L. Hart
Unilever Research Laboratories, Isleworth, Middlesex, UK

&

G. Pritchard
Kingston Polytechnic, Surrey, UK

SUMMARY

The mechanism by which fluids attack fibre-reinforced plastics is discussed. The effects of 30% sodium hydroxide solution, glacial acetic acid and 5% ferric chloride solution, on two kinds of phenolic resin reinforced with various kinds of carbon fibre, are reported and interpreted. The action of steam, air and other oxidising agents is also described. Several methods for measuring the extent of attack were assessed, and the determination of dynamic mechanical properties by means of the torsional pendulum was found to be most satisfactory.

INTRODUCTION

The term 'corrosion' is generally used in the context of metals and of particular modes of metallic degeneration. This paper follows the suggestion of Andrews[1] and takes 'corrosion' to mean any degradative process involving interaction between a material and a fluid environment (whether the interaction is chemical or physical in nature), which leads to an alteration of the structure of the material and its physical properties.

The assessment of the nature and extent of these processes in polymers is frequently carried out by methods chosen chiefly for their rapidity.

Common methods are:

(i)   measurement of changes in weight and dimensions after total immersion of small samples in various fluids
(ii) measurement of changes in flexural strength, flexural modulus and interlaminar shear strength (ILSS) after exposure
(iii) visual inspection.

To overcome the need for long exposures, the process may also be artificially accelerated by raising the temperature. This is clearly open to the objection that quite small changes in temperature can produce alterations in the structure of polymers.

Fibre-Reinforced Plastics

The mechanism of attack on homogeneous polymers, whether crystalline, amorphous or crosslinked, is broadly understood. Fibre/resin composites present a more complicated situation. They may be attacked by any of the following mechanisms:

1. swelling of the polymer matrix, leading in a few cases to dissolution, but more often resulting in small changes in volume
2. leaching of material from the matrix
3. rupture of chemical bonds in the matrix
4. swelling of the fibre
5. chemical attack on the fibre
6. destruction of the fibre–resin bond, whether by chemical or physical means.

Fibre-reinforced plastics are used increasingly in applications requiring long lifetimes. The prediction of long-term performance is therefore of great interest[2] to those concerned with material selection, both for technical and economic reasons. It is doubtful whether such predictions can be soundly based on testing programmes designed chiefly for the rapid screening of new materials.

Carbon Fibre-Reinforced Plastics

Carbon fibres have already been considered for applications involving prolonged contact with various fluids.[3] It has always been recognised that the non-availability of suitably inert resins was a major drawback. This may be overcome within a few years. The suitability of carbon fibres as reinforcements in applications requiring resistance to acids, alkalis, oil, solvents, oxidising agents and halogens has been investigated by Judd[4, 5] and by Hart and Pritchard.[6,7] The effect of certain fluids used in aircraft (fuels, brake fluids, etc.) has also been reported.[8] These results were encouraging.

The chemical resistance of carbon fibres and their composites, including those using non-resin matrices, has been reviewed by Pritchard.[9] It is found that carbon filaments undergo two main degradative processes:

1. oxidation
2. penetration, by molecules of the attacking reagent, between the carbon layer planes, with consequent swelling. This is sometimes referred to as intercalation, although the formation of stoichiometric compounds does not necessarily occur.

Resistance to oxidation increases as the maximum fibre processing temperature or heat treatment temperature (HTT) increases, but penetration intercalation shows the opposite relationship.[6] These characteristics of filament corrosion are not necessarily reflected in the behaviour of fibre-reinforced polymers.

Assessment of Corrosion

It was thought appropriate that conventional methods of assessment of deterioration should be supplemented by techniques more sensitive to matrix swelling and to interfacial bond failure. A combination of several diagnostic methods was adopted. Attention was also directed, not only to the extent of change of mechanical properties, but to the mode of failure, where this occurred.

EXPERIMENTAL

Three kinds of carbon fibre were used, distinguished by the characteristics shown in Table 1.

TABLE 1

Fibre type	Nominal modulus (GNm^{-2})	Nominal tensile strength (MNm^{-2})	HTT $(°C)$
1	350–410	1700–2200	> 2000
2	240–290	2400–3200	1600 approx.
3	190–240	1900–2600	1100 approx.

TABLE 2
LIST OF FLUIDS AND EXPOSED COMPOSITES
(A) Friedel–Crafts resin matrix

Reagent	Temp. (°C)	Untreated fibre type 1	2	3	Treated fibre type 1	2
Thirty percent w/v sodium hydroxide (100 days)	50	✓	✓	✓	✓	✓
	90	✓	✓	✓	✓	✓
Glacial acetic acid (100 days)	23	✓	✓	✓	✓	✓
	50	✓	✓	✓	—	—
Five percent ferric chloride (100 days)	23	✓	✓	✓	—	—
	50	✓	✓	✓	—	—
Five percent ferric chloride (20 days)	90	✓	✓	✓	—	—
Steam (not air-free) (50 days)	223	✓	✓	✓	—	—
Nitric acid (sp. gr. 1·21) (100 days)	23	✓	✓	✓	✓	✓
	50	✓	✓	✓	✓	✓
Air (50 days)	223	✓	✓	✓	✓	✓

(B) Phenol-formaldehyde resin matrix

Reagent	Temp. (°C)	Untreated fibre type 1	2	3
Thirty percent w/v sodium hydroxide (100 days)	50	✓	✓	✓
Glacial acetic acid (100 days)	23	✓	✓	✓
	50	✓	✓	✓
Five percent ferric chloride (100 days)	23	—	—	✓
	50	✓	✓	✓
Five percent ferric chloride (20 days)	90	—	—	✓
Nitric acid (sp. gr. 1·21) (100 days)	50	✓	✓	✓
Nitric acid (sp. gr. 1·42) (100 days)	23	✓	✓	✓
Sodium hypochlorite solu. (14% active chlorine) (100 days)	23	✓	✓	✓
Ten percent sulphuric acid (100 days)	23	✓	✓	✓
	50	✓	✓	✓

Two resins were used:

(a) a conventional phenol-formaldehyde novolak, VW 61933 (ex BXL)
(b) a commercial Friedel–Crafts resin (Xylok 210, ex Albright & Wilson Ltd.)

Unidirectional composite panels were hot-pressed in matched metal moulds of dimensions 300 mm × 85 mm to produce panels 2·5 mm thick containing 50 % w/w fibre. The panels were sawn into test bars for immersion in the fluids listed in Table 2. Exposure times are also given in Table 2.

Weight and volume changes were recorded, together with changes in flexural strength (using 40:1 span:thickness ratio). The interlaminar shear strength (5:1 span:thickness ratio) was also recorded. Both flexural strength and interlaminar shear strength were measured by means of a Howden machine, using leading nose and support diameters of 6 mm.

Flexural modulus measurements were carried out at a span:depth ratio of 100:1, to overcome the problem of penetration of the loading nose into the bars.

The shear storage modulus and logarithmic decrement were measured at a nominal frequency of 1 Hz by means of a torsional pendulum, similar in principle to that described by Learmonth et al.[11] The cross-section of the specimens was rectangular. All measurements with exposed samples were carried out at 23° ± 2 °C. The nature and extent of damage by the fluids was observed by optical microscopy and also by using a Cambridge 600 Stereoscan electron microscope.

RESULTS AND DISCUSSION

The results in Tables 3–7 are averages of four measurements.

The chemical resistance of the Friedel–Crafts resin has been described elsewhere,[10] and that of phenol-formaldehyde resins is well known. The behaviour of the isolated carbon filaments has also been established.[4, 6] The extent of damage of the carbon fibre-reinforced plastics could, therefore, be predicted in general terms. The unanswered questions are:

1. What is the relationship between fibre HTT and composite failure?
2. Which assessment method gives the most information?

(1) Thirty Percent w/v Sodium Hydroxide (Table 3)
Sodium hydroxide is perhaps the most predictable in its effects. From a

TABLE 3

CHANGES PRODUCED BY EXPOSURE TO 30% SODIUM HYDROXIDE SOLUTION FOR 100 DAYS

Property	Units	Exposure temp. (°C)	Phenolic resin Untreated fibre type			Friedel–Crafts resin Untreated fibre type			Surface treated fibres	
			1	2	3	1	2	3	2	1
Weight	Percent increase	50	55	60	60	6	10	18	—	—
Weight	Percent increase	90	—	—	—	10	16	18	—	—
Volume	Percent increase	50	300	300	300	8	6	16	—	—
Volume	Percent increase	90	—	—	—	8	17	18	—	—
Flexural strength	Percent retention	50	—	—	—	59	55	48	40	45
ILSS	Percent retention	50	×	×	×	82	77	71	57	69
ILSS	Percent retention	90	—	—	—	72	69	67	54	60
Shear modulus	Percent retention	50	—	—	—	41·0	35·8	37·5	—	—
Shear modulus	Percent retention	90	—	—	—	38·5	30·8	31·8	—	—
Log decrement	Percent increase	50	—	—	—	110	201	300	—	—
Log decrement	Percent increase	90	—	—	—	31	73	253	—	—

× = Measurement not possible.

knowledge of the structure of both resins, it is believed that sodium hydroxide attacks their hydroxyl groups and this causes gross swelling. The fibres are not attacked. Almost regardless of HTT, the ILSS falls to similar low values. The Friedel–Crafts resin contains proportionately fewer hydroxyl groups, and swelling is confined to a maximum of 18 % weight increase, 18 % volume increase, while the orthodox phenolic composite shows 60 % weight increase and up to 300 % volume increase (see Table 3). The ILSS failure mode changes from simple shear, before exposure, to homogeneous shear after immersion (possible modes of failure[12] are indicated in Fig. 1). The changes in appearance are so great that the extent of deterioration is self-evident from visual observation.

(2) Glacial Acetic Acid (Table 4)

The effects produced by immersion in glacial acetic acid are much more complex. The Friedel–Crafts resin itself is not greatly affected[10] by glacial acetic acid at 50 °C, and no significant change in weight or dimensions is found in composites with any fibre type after ambient temperature exposures, nor after exposure of the type 1 composites at 50 °C. There are small changes in the type 2 composites, and rather larger ones in type 3 specimens, after exposure at 50 °C. Yet in all the 50 °C cases, the ILSS falls dramatically; the change ranges from a 36 % decrease for type 1 to 47 % for type 3. These falls are larger than those produced by sodium hydroxide. (The phenolic composites are much more resistant.)

ILSS failure generally occurred by discrete shear, after exposure at ambient temperatures, but by homogeneous shear after immersion at 50 °C. Homogeneous shear suggests extensive breakdown of the interfacial bond. This is frequently caused by swelling, but any mechanism leading to the relief of compressive stress at the interface is likely to have this result. Acetic acid is strongly polar, and may be chemisorbed onto the fibres, to an extent dependent on the number of polar groups on the fibre surfaces. The number of fibre polar groups is believed to be related to the number of basal plane edges[13] and hence to HTT. The basal plane edges have been estimated to be in the ratio of 600:1 for types 2 and 1 fibres, respectively.[14] The chemisorption process would be temperature dependent, because of the relationship between temperature and the rate of diffusion of the acid to the fibre surfaces. It would also depend on the number of polar (hydroxyl) groups in the resin matrix.

The significance of shear modulus and damping measurements may now be considered. There has been considerable interest in the torsional properties of carbon fibre-reinforced plastics. Adams et al.[15] measured the

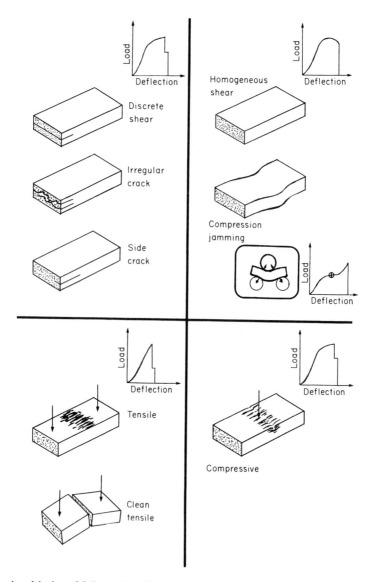

FIG. 1.   Modes of failure of unidirectionally reinforced fibre/resin composites in the interlaminar shear test. (From Ref. 12.)

TABLE 4
CHANGES PRODUCED BY EXPOSURE TO GLACIAL ACETIC ACID FOR 100 DAYS

Property	Units	Exposure temp. °C	Phenolic resin Untreated fibre type			Friedel–Crafts resin Untreated fibre type			Surface treated fibre type	
			1	2	3	1	2	3	2	1
Weight	Percent increase	23	0	0	0	0	0	0	—	—
Weight	Percent increase	50	0	0	0	0	3	9	—	—
Volume	Percent increase	23	0	0	0	0	0	0	—	—
Volume	Percent increase	50	0	0	0	0	5	16	—	—
Flexural strength	Percent retention	23	—	—	—	100	100	100	100	100
Flexural strength	Percent retention	50	—	—	—	50	46	36	—	—
ILSS	Percent retention	23	100	100	95	100	100	100	—	—
ILSS	Percent retention	50	100	97	88	64	60	53	—	—
Shear modulus	Percent retention	23	—	—	—	97·5	94·0	87·8	—	—
Shear modulus	Percent retention	50	—	—	—	25·6	25·0	18·8	—	—
Log decrement	Percent increase	23	—	—	—	39·4	30	110	—	—
Log decrement	Percent increase	50	—	—	—	155	148	453	—	—

TABLE 5

CHANGES PRODUCED BY EXPOSURE TO 5% FERRIC CHLORIDE SOLUTION FOR 100 DAYS [a]

Property	Units	Exposure temp. (°C)	Phenolic resin Untreated fibre type			Friedel–Crafts resin Untreated fibre type		
			1	2	3	1	2	3
Weight	Percent increase	23	—	—	−2	—	—	1
Weight	Percent increase	50	0	0	0	10	0	1
Weight	Percent increase	90[a]	—	—	—	0	0	0
Volume	Percent increase	23	0	0	0	0	0	0
Volume	Percent increase	50	—	—	0	0	0	0
Volume	Percent increase	90[a]	—	—	—	—	—	0
Flexural strength	Percent retention	23	—	—	—	—	—	96
Flexural strength	Percent retention	50	—	—	—	82	85	84
ILSS	Percent retention	23	—	—	100	—	—	100
ILSS	Percent retention	50	92	100	85	80	100	85
ILSS	Percent retention	90[a]	—	—	86	100	100	80
Shear modulus	Percent retention	23	—	—	—	92	—	—
Shear modulus	Percent retention	50	—	—	—	72·9	94	98·5
Shear modulus	Percent retention	90	—	—	—	94·8	85·4	97·0
Log decrement	Percent increase	23	—	—	—	15·8	—	—
Log decrement	Percent increase	50	—	—	—	15·8	11·4	−65
Log decrement	Percent increase	90	—	—	—	19·3	5·6	−12·5

[a] 90°C exposures were for 20 days.

shear modulus and specific damping capacity of polyester/carbon composites and obtained values with which the present findings may readily be reconciled. They list the possible sources of energy dissipation in fibre-reinforced plastics as:

(i) the fibres, in which very little microplastic strain can occur, and which may be regarded as making only a negligible contribution (the damping characteristics of carbon filaments have since been measured[16])

(ii) the resin matrix

(iii) the fibre–resin interface

(iv) cracks.

Cracked materials show a slight reduction in shear modulus and a large increase in damping. Damping may also be attributed to the ingress of fluid, which is not removed after the exposure. From these considerations, it is not surprising that changes in logarithmic decrement are extremely large, and far more easily detected than the changes in any other property. Moreover, changes in shear modulus are greater than changes in flexural strength. Shear modulus and damping measurements are non-destructive and can be monitored at various stages of the immersion period, using the same batch of specimens. Microcracks can be seen in the surface of Friedel–Crafts type 3 specimens removed from glacial acetic acid after immersion at 50 °C. These cracks should contribute to the development of an increased logarithmic decrement. They may also indicate the leaching of some non-bound matter from the material, as leaching in crosslinked resins is generally followed by shrinkage and the development of cracks.[17] (This helps to explain why weight changes are not very great; the increase in weight caused by fluid uptake would be partly offset by loss of other matter, as in acetone extraction procedures.) Figure 2 shows how a resin immersed in water at 70 °C first shows an increase in weight and later a decrease.

(3) Fifty Percent Ferric Chloride Solution (Table 5)

Ferric chloride solution intercalates with bulk graphite and produces slight swelling of type 1 fibres, i.e. the most graphitic kind. The action of ferric chloride on the resins themselves has not been ascertained, but might be expected to be slow. The total effect on type 2 and type 3 composites should therefore be small, and Table 5 confirms that the changes in weight and volume are not very significant. A small fall in the flexural strength of the Friedel–Crafts composites was found after exposure at 50 °C. (The

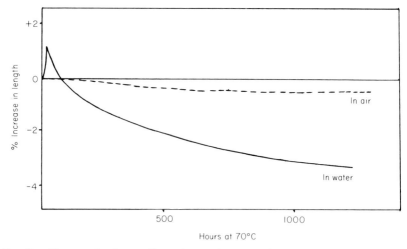

FIG. 2. Changes in linear dimensions of a cast thermosetting resin during prolonged water immersion at 70 °C, and also during heating in air at 70 °C.[17]

conventional phenolics were not examined in this respect.) The interlaminar shear strength of both types of composite also fell slightly.

Shear moduli were reduced by approximately 10 %, but the changes in logarithmic decrement were rather larger and in two cases fell instead of showing the usual rise. Both these anomalies were in type 3—Friedel–Crafts composites exposed at elevated temperatures. This particular case may be complicated by hydrolysis of the reagent itself at the highest exposure temperature.

The mechanical properties measured by the torsional pendulum show clearer evidence that some interaction has occurred between composite and fluid than any found from static mechanical property changes or weight and dimensional changes.

(4) Steam (Table 6)

Exposure to steam at 500 K (the limit of the thermal stability of the Friedel Crafts resin) for 50 days resulted in effects which are difficult to interpret unambiguously, because the steam may have contained some air. The flexural strength fell considerably more than the interlaminar shear strength. The latter test resulted in compressive failure modes. Damping increased by up to 50 % and the shear modulus fell (see Table 6). These results show a close correspondence between the change in shear modulus and the change in ILSS. Normally, the trend is similar for the two

TABLE 6
CHANGES PRODUCED IN THE FRIEDEL–CRAFTS COMPOSITES AFTER EXPOSURE TO STEAM
(NOT COMPLETELY AIR-FREE) FOR 50 DAYS AT 223 °C

Property	Units	Untreated fibre type		
		1	2	3
Weight	Percent increase	− 1	− 2	0
Volume	Percent increase	+ 1	0	+ 3
Flexural strength	Percent retention	41	44	64
ILSS	Percent retention	82	84	90
Sheer modulus	Percent retention	85	84	90
Log decrement	Percent increase	23	55	55

properties, but the magnitude of the change in shear modulus is considerably greater.

Effect of Fibre Surface Treatment

The type 3 fibres are normally sold without being surface treated, and the results obtained for type 3 composites have therefore been derived from materials as normally used. Type 2 and type 1 fibres on the other hand are usually surface treated to improve the ILSS. Most of the fibres in this present study were not treated, because the object was to isolate the effect of HTT on composite properties in the absence of any subsequent fibre modifications. The results for type 2 and type 1 composites were not therefore applicable without qualification to the corresponding materials as normally used. However, some composites of types 2 and 1 surface-treated fibres were also exposed to certain of the reagents. It was found that surface treatment did not improve the resistance to sodium hydroxide. This reagent produced resin swelling. Surface treatment considerably improved the behaviour of the composites in nitric acid (sp. gr. 1·21), both at 23 °C and

TABLE 7(A)
FRIEDEL–CRAFTS MATRIX. EFFECT OF OXIDISING AGENTS ON PERCENT RETENTION OF ILSS

Reagent	No. of days	Temp.	Untreated fibre type			Treated fibre type	
			1	2	3	1	2
Nitric acid (sp. gr. 1·21)	100	23	70	84	98	90	96
Nitric acid (sp. gr. 1·21)	100	50	7	4	68	42	60
Air	50	223	62	64	100	91	96

TABLE 7(B)

FRIEDEL–CRAFTS MATRIX. EFFECT OF OXIDISING AGENTS ON PERCENT RETENTION OF
FLEXURAL STRENGTH

Reagent	No. of days	Temp.	Untreated fibre type			Treated fibre type	
			1	2	3	1	2
Nitric acid (sp. gr. 1·21)	100	23	82	80	94	90	96
Nitric acid (sp. gr. 1·21)	100	50	×	×	55	28	63
Air	50	223	37	40	68	62	70

× = Measurement not possible.

TABLE 7(C)

PHENOL FORMALDEHYDE MATRIX. EFFECT OF OXIDISING AGENTS ON PERCENT RETENTION
OF ILSS (100 DAYS' EXPOSURE)

Reagent	Temp. (°C)	Untreated fibre type		
		1	2	3
Nitric acid (sp. gr. 1·21)	50	×	×	×
Nitric acid (sp. gr. 1·42)	23	×	×	×
Sodium hypochlorite (14 percent active chlorine)	23	89	94	95
10 percent sulphuric acid	23	100	100	100
10 percent sulphuric acid	50	100	100	100

× = Measurement not possible.

more particularly at 50 °C, and also in air at 223 °C (50 days). This is because a strong fibre–resin bond protects the fibres against oxidation (see Table 7).

CONCLUSIONS

A study of corrosion mechanisms in carbon fibre-reinforced plastics shows that several different modes of attack are possible.

The detection of comparatively slight damage is conveniently carried out by non-destructive dynamic tests with a torsional pendulum, at approximately 1 Hz.

Surface treatment of fibres does not always improve the chemical resistance of composites, but can do so in the case of attack by oxidising agents.

ACKNOWLEDGEMENT

One of us (G. L. H.) acknowledges the provision of a studentship by the Courtaulds Educational Trust Fund.

REFERENCES

1. ANDREWS, E. H. *Fracture in polymers*, Oliver and Boyd, London (1968) p. 86.
2. PRITCHARD, G. *Reinforced plastics*, **18** (1974) p. 155.
3. HART, G. L. and PRITCHARD, G. *Proc. 2nd Intl. Carbon Fibre Conf.*, *London*, Plastics Institute (1974) Paper 39.
4. JUDD, N. C. W. *Proc. 1st Intl. Carbon Fibre Conf.*, *London*, Plastics Institute (1971) Paper 32.
5. JUDD, N. C. W. *27th Reinforced Plastics Conf.*, *Washington, DC*, SPI. (1972) Paper 3-D.
6. HART, G. L. and PRITCHARD, G. *27th Reinforced Plastics Conf.*, *Washington, DC*, SPI. (1972) Paper 3-E.
7. HART, G. L. and PRITCHARD, G. *29th Reinforced Plastics Conf.*, *Washington, DC*, SPI. (1974) Paper 2-A.
8. PRITCHARD, G., MATHEWS, B. J. and STOKES, F. C. MOD Procurement Executive Report, AT/2055/06MAT, 1975.
9. PRITCHARD, G. *Polymer-Plast. Technol.-Eng.*, **1**(5) (1975) p. 55.
10. HARRIS, G. I. *Brit. Polym. J.*, **2** (1970) p. 270.
11. LEARMONTH, G. S., PRITCHARD, G. and REINHARDT, J. *J. Appl. Polym. Sci.*, **12**(3) (1968) p. 403.
12. DANIELS, B. K., HARAKAS, N. K. and JACKSON, R. C. *Fibre Sci. Technol.*, **3**(3) (1971) p. 187.
13. BUTLER, B. L., LEMAISTRE, C. W. and DIEFENDORF, R. J. *28th Reinforced Plastics Conf.*, *Washington, DC*, SPI. (1973) Paper 21-C.
14. CLARK, D., WADSWORTH, N. J. and WATT, W. *Proc. 2nd Intl. Carbon Fibre Conf.*, *London*, Plastics Institute, (1974) Paper 7.
15. ADAMS, R. D., FOX, M. A. O., FLOOD, R. J. L., FRIEND, R. J. and HEWITT, R. L. *J. Comp. Mater.*, **3** (1969) p. 594.
16. ADAMS, R. D. and LLOYD, D. H. *J. Phys. E: Scientific Instruments*, **8** (1975) p. 475.
17. TANEJA, N. Studies of water damage in polyester glass laminates. Ph.D. Thesis, Kingston Polytechnic, 1974.

Chapter 2

RUPTURE DURING BENDING OF COMPOSITES REINFORCED WITH DISCONTINUOUS AND UNIDIRECTIONAL METAL FIBRES: STUDY BASED ON A RECORDED TEST

A. Cupcic & J. M. Berthelot

Laboratory of Physics of Materials, Le Mans University, France

SUMMARY

Composites containing discontinuous and unidirectional metal fibres were made. Their shock fracture by means of a three-point non-notched bending was studied with the help of a recorded test (Charpy impact machine with a high natural frequency dynamometrical component).

The exploitation of these recorded tests enables us to verify a scheme (fracture energy, extraction and fibre shearing energy, determination of the critical length). The scheme is obtained from several concordant cross-checks.

NOMENCLATURE

F Force exerted by the blade of the machine.

Z Parameter of the linear travel of the blade.

β Half of the opening angle of the two sides of the rupture plane.

d Diameter of the fibres.

l Length of the fibres.

l_c Critical length.

l^* $l^* = l$ if $l < l_c$ and $l^* = l_c$ if $l > l_c$.

$b \times h$ Cross-section of the test piece (here $10 \times 3 \cdot 5 \, \text{mm}^2$).

τ Coefficient of friction on extraction.

L 40 mm (distance between the two stand-pieces of the Charpy impact machine).

I. EXPERIMENTAL PART OF THE STUDY

(a) Composite
We used:

> As resin, Araldite CY 219/hardener HY 977/accelerator; DY 060 (CIBA
> company) with the following percentages, respectively: 62·50%;
> 31·25% and 6·25%.
> As fibres: fine reheated copper wires (Thompson Company) $d = 0·10$ and
> 0·20 mm, cut into lengths of 10 and 20 mm.

The composites were made as a result of successive pourings at room
temperature and pressure, the fibres being laid out unidirectionally. After a
setting time of 60 h at 50 °C, the test pieces were removed parallel to the
fibres and then machined to the dimensions of $60 \times 10 \times 3·5$ mm^3.

The proportion in volume v_j of the fibres is limited to:

$$l = 10 \text{ mm} \qquad d = 0·10 \text{ mm} \qquad v_j < 0·15$$

$$l = 10 \text{ mm} \qquad d = 0·20 \text{ mm} \qquad v_j < 0·30$$

$$l = 20 \text{ mm} \qquad d = 0·20 \text{ mm} \qquad v_j < 0·20$$

(b) Testing Equipment
We used a Charpy impact machine 5101 Zwick, the axis of which was
equipped with a rotative transducer Schaewitz R 30 C (measurement of
travels and speeds) and whose blade was altered so as to record forces by
means of the strain-gauge method.[1]

Curves are shown on a dual beam oscilloscope Philips PR 3231 and
photographed (Fig. 1).

The characteristics of the dynamometrical unit are as follows:

natural frequency of the dynamometrical component: 60 kHz,
filtering of the signal by means of a cut-off frequency low-pass filter:
   10 kHz,
sensitivity: 6N,
signal/noise ratio > 40 for a force of 200N,
linearity $< \pm 0·5\%$ between 0 and 2500N,
transmitting range at -1 dB/0–10 kHz,
response time: 50 μsec,
negligible overshoot.

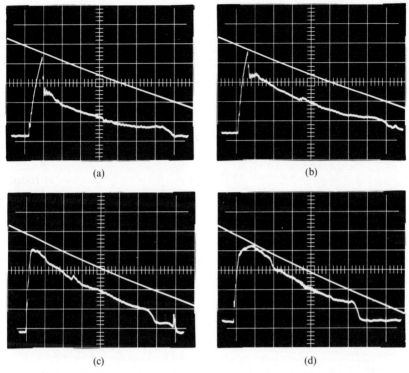

FIG. 1.   Experimental curves ($l = 10$ mm, $d = 0·20$ mm). Each upper curve is dz/dt (t) given by rotative transducer. Each lower curve is $F(t)$ given by dynamometric gauge. (a) $v_j = 0·10$; (b) $v_j = 0·15$; (c) $v_j = 0·23$; (d) $v_j = 0·30$.

(c) Shape of Force-Deflexion Curves
We obtained curves of two types (Fig. 2):

	Type (a)	Type (b)
$l = 10$ mm $d = 0·20$ mm	$v_j < 0·23$	$v_j > 0·23$
$l = 20$ mm $d = 0·20$ mm	$v_j < 0·16$	$v_j > 0·16$
$l = 10$ mm $d = 0·10$ mm	—	$v_j > 0$

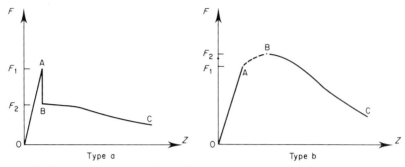

FIG. 2.   Schematic representation of the two types of experimental curve. Force F.
Deflexion Z.

It is proved in the following that:

OA is an elastic part (curves of types (a) and (b)).

Point A corresponds to the moment when the crack appears (curves of types (a) and (b)) (force F_1).

The crack ends at point B (force F_2) in a time shorter than the response time of the testing-unit (curve of type (a) with $F_2 < 1$).

Or the crack spreads progressively from A to B (end of crack) in a measurable total time (curve of type (b) with $F_2 > F_1$).

The part BC corresponds to the tearing of fibres by pulling apart the two sides of the crack (curves of types (a) and (b) with F decreasing).

In C, the test piece is propelled from the impact machine, some of the fibres remaining not completely extracted.

In addition, we will differentiate the elastic energy W_e (between O and A), the energy W_P after matrix rupture (between B and C), the total energy W_t absorbed during the shock (between O and C). It should be noted that since the extraction of the fibres is not complete in C, there remains a potential extraction energy which was not measured in this test but which can be calculated by consulting the scheme given below.

II.   SUGGESTED SCHEME

(a) Geometry of the Experiment

From O to A, there occurs an elastic deformation of flexure (Fig. 3(a)).

From A to B the crack is complete, and the two halves of the test piece have gone upwards, forming the corresponding opening angle 2β (Fig. 3(b)).

The elastic energy W_e (stored between O and A) and contingently the energy stored between A and B supplies the rupture energy of the matrix and possibly the extraction and shearing energy of the fibres that correspond to the gap (Fig. 3(b)), as well as the rupture energy of some fibres when $l > l_c$.

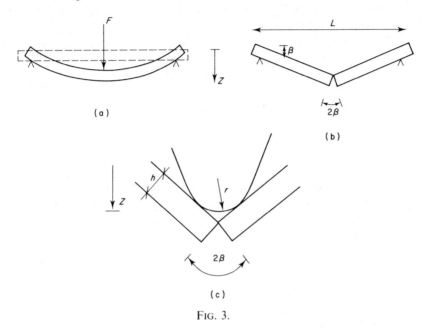

(a)

(b)

(c)

FIG. 3.

From B to C there occurs the extraction of the fibres as the angle is widened and deformation as the extracted fibres are shorn (this can be checked by examining the test pieces).

(b) Interaction Fibre–Matrix[2−5]

We assert that in the shock-rupture process, the scheme proposed by Kelly[2] is valid: an interaction fibre matrix characterised during extraction by a coefficient of friction τ. The possible dependence of τ on the speed of the test has not been taken into consideration in this test. (In our tests $dZ/dt \simeq 1\cdot7\,\mathrm{m\,sec^{-1}}$, that is an extraction speed of the fibres $<0\cdot9\,\mathrm{m\,sec^{-1}}$.) We then keep the notion of critical length

$$l_c = \frac{d}{2\tau}\,\sigma_{f\,\mathrm{max}}$$

(c) Layout of the Fibres
The fibres are laid out parallel to one another in a random manner. An elementary representation, convenient and sufficiently accurate for calculation, is as follows: fibres laid out in accordance with the height h of the test piece, in p identical and equally spaced layers. In a layer plane, the middle points of each fibre are displaced by a distance Δ parallel to the length of the test piece (Fig. 4).

The number of fibres crossing the centre plane of rupture is then:

$$N = A v_f \quad \text{and} \quad \Delta = \frac{l_p}{2N}$$

(for the calculation of A see Appendix I). If $l \leq l_c$ all the fibres are extracted without rupture. If $l \geq l_c$ fibres (in number of $(l_c/l)N$) are extracted without rupture and fibres [in number of $N(1 - l_c/l)$] are broken without being extracted.

(d) Shearing Deformation of the Fibres Extracted Without Rupture
This deformation may be observed on the test pieces after the test has been completed. We adopt the plastic shearing scheme in which the shearing stress τ_0 is constant $\sim \sigma_{f\,\text{max}}/2 = \tau l_c/d$.

III. PREDICTION OF W_P AND $F(Z)$[1]

(a) Calculation of W_P
The detailed calculations are given in Appendix II.

As shown above, the predicted value of W_P may be expressed as depending on n_t as well as on the opening angle 2β and again on the two possibilities $l < l_c$ and $l > l_c$.

We may then write the result as depending on v_f and $\alpha = h/l^* \sin \beta$. These two possibilities $l < l_c$ and $l > l_c$ are then found together in the same formula using the parameter

$$l^* = l \quad \text{if } l < l_c$$
$$l^* = l_c \quad \text{if } l > l_c$$

Then:

$$W_P = \frac{bh}{ld}(l^*)^3 \left(f(\alpha) + \frac{l_c}{h} g(\alpha) \right) \tau v_f$$

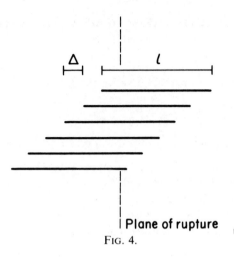

Plane of rupture

FIG. 4.

If $\alpha < \frac{1}{4}$

$$f(\alpha) = \alpha - \tfrac{8}{3}\alpha^2 + \tfrac{8}{3}\alpha^3$$

$$g(\alpha) = \alpha^2 - \tfrac{16}{9}\alpha^3$$

If $\alpha > \frac{1}{4}$

$$f(\alpha) = \tfrac{1}{6} - \tfrac{1}{96}\alpha$$

$$g(\alpha) = \tfrac{5}{144} + \tfrac{1}{24}\log 4d$$

(b) Calculation of $F(Z)$ Between B and C

The experimental results (e.g. the fact that the two halves of the test piece are straightened back to their original position) lead us to believe that the elastic energy has been reproduced between A and B.

Hence, we may write that $F(Z) = \mathrm{d}W_P/\mathrm{d}Z$ since $Z > Z_\mathrm{B}$.

The geometry of the test between B and C, proved by examining the test piece after rupture gives us (Fig. 3(c)):

$$Z = \frac{L}{2}\,tg\beta + (r + h)\left(1 - \frac{1}{\cos\beta}\right)$$

$$F(Z) = \frac{\mathrm{d}W_P}{\mathrm{d}x} \cdot \frac{\mathrm{d}\alpha}{\mathrm{d}\beta} \cdot \frac{\mathrm{d}\beta}{\mathrm{d}Z}$$

$$F(Z) = \frac{bh^2}{ld}(l^*)^2 \left(\frac{\mathrm{d}f}{\mathrm{d}\alpha} + \frac{l_c}{h} + \frac{\mathrm{d}g}{\mathrm{d}\alpha}\right)\frac{\cos^3\beta}{L/2 - (r + h)\sin\beta}\,\tau v_f$$

$$F_2 \sim \frac{bh^2}{ld}(l^*)^2 \frac{\tau v_f}{(L/2) - 2(r + h)(Z_\mathrm{B}/L)}$$

IV. AGREEMENT OF THE SCHEME WITH EXPERIMENTAL RESULTS

(a) Study of F_2

We may plot the experimental curves in the following form (Fig. 5):

$$f(v_j) = \frac{F_2}{bh^2}\left(\frac{L}{2} - 2(r + h)\frac{Z_B}{L}\right)$$

The experimental points are grouped in two straight lines:

(i)   One straight line D_1 whose slope 170 MPa corresponds to l = 20 mm, $d = 0.20$ mm and $l = 10$ mm, $d = 0.10$ mm. (D_1 corresponds to the case when $l > l_c$).

(ii)   One straight line D_2 whose slope 122 MPa corresponds to l = 10 mm, $d = 0.20$ mm. (D_2 corresponds to the case when $l < l_c$).

There is a concordance with the theoretical prediction:

$$f(v_j) = (dl)^{-1}(l^*)^2\tau v_j$$

Hence, we may deduce the following numerical values:

$$\tau = 2.28 \text{ MPa} \qquad \text{and} \qquad r_c = \frac{l_c}{d} = 80$$

to which corresponds the value $\sigma_{f\,\max} = 2\tau r_c \sim 365$ MPa.

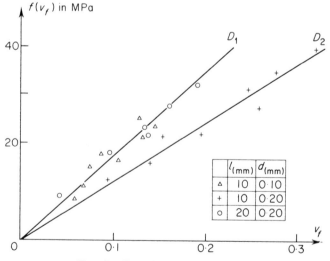

FIG. 5.   Experimental values of $f(v_j)$.

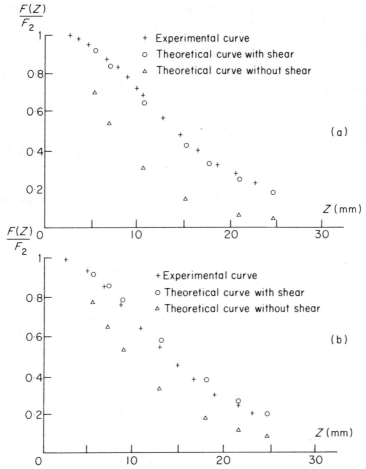

Fig. 6. (a) Case $l = 10$ mm, $d = 0\cdot20$ mm (here $l < l_c$). (b) Case $l = 20$ mm, $d = 0\cdot20$ mm (here $l > l_c$).

The experimental value of $\sigma_{f\,max}$ measured directly on the continuous fibre under bending test conditions is $\sigma_{f\,max} > 300$ MPa.

(b) Study $F(Z)$ Between B and C

We plotted three curves $F(Z)$ carried back to the same original value F_2 in B.

Thus for $l = 10$ mm, $d = 0\cdot20$ mm ($l < l_c = 16$ mm from above) (Fig. 6(a)) and $l = 20$ mm; $d = 0\cdot20$ mm ($l > l_c$) (Fig. 6(b)): the three curves are first the experimental curve, secondly the curve representing the above-mentioned theoretical formula and, thirdly, the curve representing the above-mentioned theoretical formula but without considering the shearing energy of the fibres. The concordance of the first two curves justifies the total validity of the proposed scheme and in particular the fact that the shearing energy must be here taken into account.

The plotting of the experimental curves does not depend on v_f. The plotting of the curve $F(Z)/F_2$ may be used to calculate in a different way the value of l_C which appears in the coefficient of $(dg/d\alpha)$. This new means of calculation is identical to that we adopt below in the study of W_P.

(c) Study of W_P

From the recorded experimental curves $F(t)$ and $Z(t)$, we may rearrange the curves $F(Z)$ so as to deduce

$$W_P = \int_B^C F(Z)\,dZ$$

that is $W_P(v_f)$ (Fig. 7).

Moreover, we may calculate W_P from the theoretical formula found in Section III(a); Z_B (value found at the end of the rupture process of the matrix) corresponds roughly to 2·5 mm, Z_C corresponds to ~ 25 mm (test piece is expelled from the supports). By comparing the theoretical and

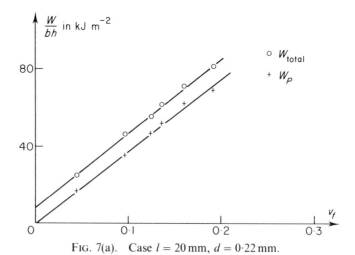

FIG. 7(a).   Case $l = 20$ mm, $d = 0.22$ mm.

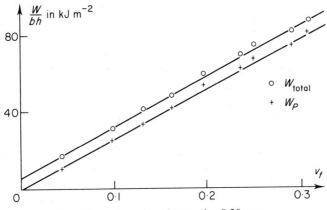

FIG. 7(b). Case $l = 10$ mm, $d = 0·20$ mm.

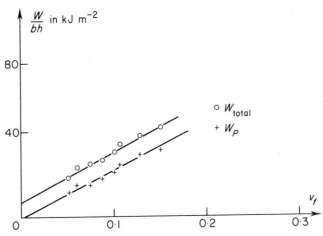

FIG. 7(c). Case $l = 10$ mm, $d = 0·10$ mm.

experimental values of W_P we obtain a new measurement of τ. Thus we have:

$l\,mm$	$d\,mm$	$W_P(bh)^{-1}(v_f)^{-1}$ in kJm^{-2}	τ in MPa
10	0·20	262	2·36
20	0·20	383	2·38
10	0·10	240	2·31

Once more, we find a new concordance of the value of τ.

In addition, we can deduce from our tests that the energy absorbed in the shearing of the fibres represents about half W_p. and this, among other possible meanings, implies that any calculation from an integral measurement of W_p and from a single friction scheme would give for τ a value approximately double the real value.

V. CONCLUSION

Further information (for instance energy in creating the crack, propagation energy of the crack up to total rupture, etc.) may be deduced from the exploitation of our recordings and will be dealt with later on, as will the technological aspect of the measurements bearing on the introduction of copper fibres in the epoxy resin (increase of W_t, etc.).

The rupture energy may always be known by means of an integral measurement. We have here demonstrated that the recorded test may, alone, allow us to relate this energy to a specific scheme on metal fibred composites (fibres capable of plastic deformation). The recorded test alone may also enable us to determine an accurate value of τ and therefore of l_c and finally to compare rupture tests by bending, using different geometries.

APPENDIX I

All cross-sections being identical we thus obtain:

$$N \frac{\pi d^2}{4} = bh \cdot v_J$$

N being the number of fibres crossing the section $b \times h$. For the calculations in Appendix II, these N fibres are supposed to be displayed according p identical layers, each containing N/p fibres.

In one layer, the displacing Δ (Fig. 4) is then

$$\Delta = \frac{l}{2} \cdot \frac{p}{N}$$

APPENDIX II

(1) Energy of Extraction Without Shearing in the Case when $l < l_c$

(a) *Preliminary problem* (Fig. 8): We separate from x the two faces of the

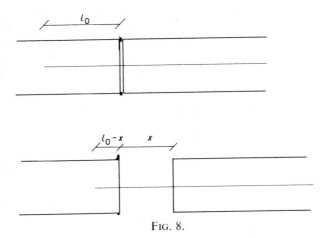

FIG. 8.

rupture plan, the initial length of the fibre from the extracted side being l_0. The friction force being $\pi d\tau$ per unit of length of fibre, the energy of extraction for a single fibre is:

$$W_1(x, l_0) = \frac{\pi d\tau}{2}(2l_0 x - x^2) \qquad \text{if } x < l_0$$

$$W_1(x, l_0) = \frac{\pi d\tau}{2} l_0^2 \qquad\qquad \text{if } x > l_0 \qquad\qquad \text{(a)}$$

(the fibre being completely extracted). For the whole of Appendix II, we shall always have to consider both possibilities: the fibres are or are not completely extracted.

(b) *Calculation of W_{ext}*: In accordance with the proposed scheme: p identical layers correspond to the dimension $y_j = (h/2p)(2j - 1)$ for layer No. j $(1 \leq j \leq p)$. Each layer contains N/p fibres whose total lengths which have to be extracted are, respectively:

$$l_0 = \Delta/2;\ 3\Delta/2, \ldots;\ (2k - 1)\Delta/2; \ldots;\ (2N/p - 1)\Delta/2 \qquad (1 \leq k \leq N/p)$$

Figure 9 establishes the correspondence between x, length previously extracted and β, the folding angle, that is $x = 2y_j \sin \beta$.

A fibre may then be spotted with the two marks (j, k); hence a representation of each fibre by a point in the orthogonal co-ordinates (j, k). The whole of the fibres is represented in the form of a squaring of the rectangle. $(1 \leq j \leq p;\ 1 \leq k \leq N/p)$, each fibre corresponding to an integer value of j and k (Fig. 10).

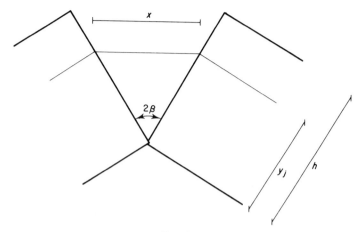

FIG. 9.

For each fibre we have:

$$x = \frac{h}{p}(2j - 1)\sin \beta$$

$$l_0 = (2k - 1)\Delta/2 = (2k - 1)\frac{l_p}{4N} \qquad \text{(b)}$$

In the co-ordinates (j, k), the condition $x = l_0$ (complete extraction of a fibre) may be written, putting down:

$$\alpha = \frac{h}{l}\sin \beta$$

$$x = l_0 \rightarrow (2k - 1) = \frac{4hN}{l_p^2}(2j - 1)\sin \beta = \frac{4\alpha N}{p^2}(2j - 1)$$

This is the equation of the straight line $D(\alpha)$ which splits, in co-ordinates j, k, the space of fibres into two parts: the part (D_1) containing the fibres totally extracted, the part (D_2) containing the fibres not totally extracted; this for a value of β, that is of α fixed (Fig. 10).

Therefore, the total energy of extraction of all the fibres is, for a value of α, given by the sum of a double series:

$$W_{\text{ext}} = \sum\sum_{(D_1)} \frac{\pi d\tau}{2}l_0^2 + \sum\sum_{(D_2)} \frac{\pi d\tau}{2}(2l_0 x - x^2) \qquad \text{(c)}$$

Where x and l_0 are to be replaced depending on the marks j, k of each fibre according to eqn. (a).

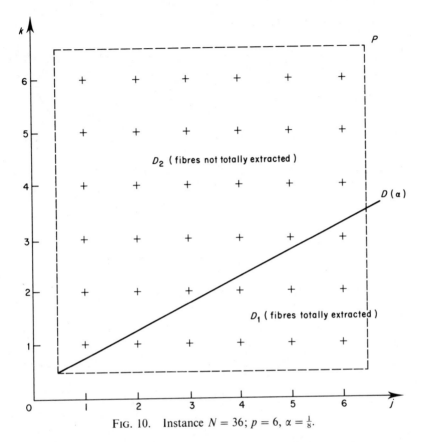

FIG. 10. Instance $N = 36$; $p = 6$, $\alpha = \frac{1}{8}$.

The direct summation of this double series has been given in reference 1. As a matter of fact, this calculation, as soon as p (and therefore N) goes beyond a few units, may be regarded as equal to that of the double integral extended to (D_1) and (D_2):

$$W_{\text{ext}} = \iint_{(D_1 + D_2)} W_1(j, k)\, dj\, dk \qquad \text{(d)}$$

each fibre being the centre of a unit square in (j, k) co-ordinates, and taking for summation

$$\tfrac{1}{2} \le j \le p + \tfrac{1}{2}$$

$$\frac{1}{2} \le k \le \frac{N}{p} + \frac{1}{2}$$

we can write $W_1(j, k)$ deduced from (a) and (b):

$$W_1(j, k) = \frac{\pi d\tau}{2} l_0^2 = \frac{\pi d\tau}{8} \Delta^2 (2j - 1)^2 = \frac{\pi d\tau}{32 N^2} l^2 p^2 (2j - 1)^2 \qquad \text{in } D_1$$

$$W_1(j, k) = \frac{\pi d\tau}{2} (2l_0 x - x^2) = \frac{\pi d\tau}{4} \frac{l^2}{N}$$

$$\times \alpha \left((2j - 1)(2k - 1) - \frac{2\alpha N}{p^2} (2j - 1)^2 \right) \qquad \text{in } D_2$$

The calculation of eqn. (d) is elementary. According to the relative position of $D(\alpha)$ and thus of the point P ($j = P + \frac{1}{2}; k = N/p + \frac{1}{2}$) (Fig. 10), that is to say according to the respective value of α and of $\frac{1}{4}$, we find:

If:

$$\alpha = \frac{h}{l} \sin \beta \leq \frac{1}{4}$$

$$W_{\text{ext}} = \pi d\tau l^2 N \left(\frac{\alpha}{4} - \frac{2\alpha^2}{3} + \frac{2}{3}\alpha^2 \right)$$

and with:

$$N = \frac{4}{\pi d^2} bh v_J$$

$$W_{\text{ext}} = \frac{bhl^2}{d} \tau \left(\alpha - \frac{8\alpha^2}{3} + \frac{8\alpha^3}{3} \right) v_J$$

If:

$$\alpha \geq \frac{1}{4}$$

$$W_{\text{ext}} = \pi d\tau l^2 N (\tfrac{1}{24} - \tfrac{1}{384}\alpha) = \frac{bhl^2}{d} \tau (\tfrac{1}{6} - \tfrac{1}{96}\alpha) v_J$$

Of course, we have $\sin \beta \leq 1 \to \alpha \leq h/l$.

(c) *Energy of extraction without shearing when $l > l_c$:* The fibres, $N(1 - l_c/l)$ in number, for which $l > l_c$ are broken and their rupture energy is not included in the calculation of W_{ext}.

We only have to calculate the energy of extraction of the fibres for which $l \leq l_c$, whose number is $N(l_c/l)$ and whose spacing is:

$$\Delta = \frac{l}{2} \frac{p}{N} = \frac{l_c}{2} \frac{p}{N(l_c/l)}$$

We may see that the above results are worth considering in the case of the parameter l^* ($l^* = l$ if $l < l_c$ and $l^* = l_c$ if $l > l_c$). The number of fibres extracted is always $N(l^*/l)$, the length of the fibres is l^*, their spacing is $\Delta = l_p/2N$, and putting $\alpha = (h/l^*) \sin \beta$, we have the formula:

$$W_{ext} = \frac{bh(l^*)^3}{ld} \tau(\alpha - \tfrac{8}{3}\alpha^2 + \tfrac{8}{3}\alpha^3)v_f \qquad \text{if } \alpha \leq \tfrac{1}{4}$$

$$W_{ext} = \frac{bh(l^*)^3}{ld} \tau(\tfrac{1}{6} - \tfrac{1}{96}\alpha)v_f \qquad \text{if } \alpha \geq \tfrac{1}{4}$$

thus in the two cases $l \leq l_c$ or $l \geq l_c$.

(2) Calculation of the Shearing Energy

If we only take into consideration the energy of extraction previously calculated, it is not possible to obtain a satisfactory concordance between theoretical and experimental results.

For instance, the experimental results lead to a value of τ of $2\cdot3$ MPa obtained from F_2 and a value of 4 MPa obtained from values of W_p. In addition, the theoretical curve of the force applied on the blade function of its movement shows a more rapid decrease than the experimental curve (Fig. 6).

Also, we may see, after experimentation, that part of the extracted fibres is straight. It is then necessary to consider that a part of the energy absorbed during the plastic process comes from the shearing of the fibres.

The results obtained show the validity of such a scheme in the case of fibres made of reheated copper.

The adopted scheme is as follows: the experimental curve for the shearing of copper is regarded as made up only of a plastic part with a constant stress τ_0.

If l_0 was the initial length to be extracted and x the extracted length the shearing energy of a fibre is (Fig. 9):

$$W_1 = \frac{\pi d^2}{4} \tau_0 \times \sin \beta \qquad \text{if } x < l_0$$

and with again, for the fibre marked (j, k):

$$x = \frac{h}{p}(2j - 1)\sin \beta = \frac{\alpha l}{p}(2j - 1)$$

$$l_0 = (2k - 1)\frac{\Delta}{2} = (2k - 1)\frac{lp}{4N}$$

with again $\alpha = (h/l)\sin\beta$

$$W_1 = \frac{\pi d^2}{4}\tau_0\frac{\alpha^2 l^2}{ph}(2j-1) \qquad \text{if } x < l_0$$

(fibre not totally extracted).

The fibre is totally extracted when

$$x = l_0 \qquad \text{or} \qquad \alpha = \frac{(2k-1)}{(2j-1)}\cdot\frac{p^2}{4N}$$

then the constant value of W_1 is:

$$W_1 = \frac{\pi d^2}{4}\tau_0\frac{l^2 p^3}{16hN^2}\frac{(2k-1)^2}{(2j-1)} \qquad \text{if } x > l_0$$

(fibre totally extracted).

The process of calculation will be identical to that described above with the function of the shearing energy of a fibre:

$$W_1(j,k) = \frac{\pi d^2}{64}\tau_0\frac{l^2 p^3}{hN^2}\frac{(2k-1)^2}{(2j-1)} \qquad \text{in } D_1$$

$$W_1(j,k) = \frac{\pi d^2}{4}\tau_0\frac{\alpha^2 l^2}{ph}(2j-1) \qquad \text{in } D_2$$

When $l < l_c$ we found:

$$W_{\text{cis}} = \frac{\pi d^2\tau_0 l^2 N}{h}\left(\frac{\alpha^2}{4}-\frac{4\alpha^3}{9}\right) = \tau_0 l^2\, b\left(\alpha^2-\frac{16\alpha^3}{9}\right)v_f$$

if: $$\alpha < \tfrac{1}{4}$$

and: $$W_{\text{cis}} = \frac{\pi d^2\tau_0 l^2 N}{96h}(\tfrac{5}{6}+\log 4\alpha) = \tau_0 l^2 b(\tfrac{5}{144}+\tfrac{1}{24}\log 4\alpha)v_f$$

if: $$\alpha > \tfrac{1}{4}$$

There again, both possibilities $l < l_c$ and $l > l_c$ combine into one single formula using l^* and $\alpha = (h/l^*)\sin\beta$:

$$W_{\text{cis}} = \frac{\tau_0(l^*)^3 b}{l}\left(\alpha^2-\frac{16\alpha^3}{9}\right)v_f \qquad \text{if } \alpha < \tfrac{1}{4}$$

$$W_{\text{cis}} = \frac{\tau_0(l^*)^3 b}{l}(\tfrac{5}{144}+\tfrac{1}{24}\log 4\alpha)v_f \qquad \text{if } \alpha > \tfrac{1}{4}$$

Then using the classical formula in metallurgy $\tau_0 \sim \sigma_{max}/2$ and with σ_{max} = $2\tau l_c/d$ hence $\tau_0 = \tau l_c/d$ we may express the shearing energy:

$$W_{cis} = \frac{b(l^*)^3 \tau l_C}{ld}\left(\alpha^2 - \frac{16\alpha^3}{9}\right)v_f \qquad \text{if } \alpha < \tfrac{1}{4}$$

$$W_{cis} = \frac{b(l^*)^3 \tau l_C}{ld}(\tfrac{5}{144} + \tfrac{1}{24}\log 4\alpha)v_f \qquad \text{if } \alpha > \tfrac{1}{4}$$

with $\alpha \le h/l^*$.

REFERENCES

1. BERTHELOT, J. M., Recherche de l'obtention de matériaux composites résines–fibre métallique à grande énergie de rupture. Thèse 3° cycle, Poitiers, 1974.
2. KELLY, A., *Strong solids*, Clarendon Press, 1966.
3. COTTREL, A. H., *Proc. Roy. Soc.*, **A 282** (1964) p. 2.
4. KELLY, A., *Proc. Roy. Soc.*, **A319** (1970) p. 95.
5. HELFET, J. L. and HARRIS, S., Fracture toughness of composites reinforced with discontinuous fibres, *J. Mat. Sci.*, **7** (1972) pp. 494–8.

Chapter 3

VIBRATIONS OF ANTISYMMETRICALLY LAMINATED CYLINDRICAL SHELLS

Y. V. K. SADASIVA RAO & P. C. RAJU
Vikram Sarabhai Space Centre, Trivandrum, India

SUMMARY

The recent development of high-modulus fibres has resulted in lightweight, high-strength structures which find applications in the aerospace and related industries. The availability of these materials has in turn encouraged the development of theoretical models capable of describing composite material behaviour under static and dynamic loads. One of such structural elements is the cross-ply laminated circular cylindrical shell.

The lamination asymmetries about the middle surface of the shell cause coupling between bending and extension of the laminate. This is an important step in learning how to properly laminate the shell. The object of this paper is to present a theory with numerical results for the vibration of cross-ply laminated circular cylindrical shells under simply supported boundary conditions. The behaviour of laminates that are antisymmetric has been studied using a modified shell theory. The results are compared with exact solution and an earlier work which uses Donnell's approximation. The effect of coupling between bending and extension has been studied in detail varying the ratios of length to radius and radius to thickness of the cylinder.

NOMENCLATURE

A_{ij}	Extensional stiffness of a laminated shell.
B_{ij}	Coupling stiffness of a laminated shell.
D_{ij}	Bending stiffness of a laminated shell.
E_1	Young's modulus in the 1-direction of a lamina.

E_2	Young's modulus in the 2-direction of a lamina.
G_{12}	Shear modulus in the 1-2 plane of a lamina.
K_ω	Normalised frequency.
L	Length of the shell.
$M_x, M_\theta, M_{x\theta}$	Stress couples.
m	Number of half-waves in the x-direction.
$N_x, N_\theta, N_{x\theta}$	Membrane stress resultants.
n	Number of half-waves in θ-direction.
$\bar{n}$	Number of layers.
Q_x, Q_θ	Shearing stress resultants.
R	Radius of the shell to the middle surface.
t	Thickness of the shell.
u, v, w	Displacement components in x-, θ-, and $\bar{z}$-directions, respectively.
x, θ	Shell co-ordinates.
$\varepsilon_x, \varepsilon_\theta, \varepsilon_{x\theta}$	Membrane strains.
$\kappa_x, \kappa_\theta, \kappa_{x\theta}$	Change of curvature.
v_{12}, v_{21}	Poisson's ratio.
ρ	Mass density of the shell.
ω	Circular frequency in rad/sec.
$'$	Differentiation with respect to x.
$*$	Differentiation with respect to θ.

INTRODUCTION

The geometry of the shell under consideration is shown in Fig. 1, in which the fibres in one layer are in the axial direction and those in the next layer are in the circumferential direction. The lamination asymmetries about the shell middle surface cause coupling between bending and extension of the laminate. This phenomenon is evidenced by bending of a laminate that is subjected to only in-plane or in-surface forces or extension of a laminate that is bent by the application of moments alone. It has been shown by Whitney and Leissa[1] that this coupling reduces the frequencies in the case of laminated plates. Jones[2] has presented an exact solution for simply supported plates that are laminated unsymmetrically about their middle surfaces. It is shown that, for the case of antisymmetric laminates, the effect of coupling between bending and extension dies out rapidly as the number of layers is increased. The same theory has been extended to simply supported circular cylindrical shells[3] wherein Donnell's approximation is used.

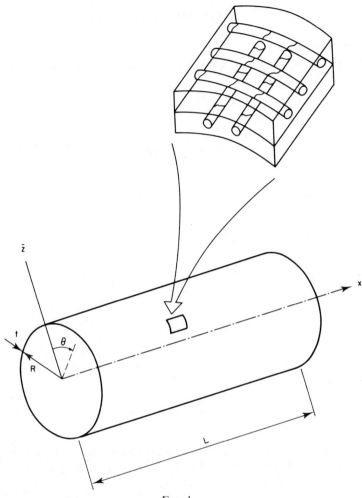

FIG. 1.

It is shown in a previous paper[4] that by assuming the cylinder to be inextensible in the circumferential direction (i.e. $\varepsilon_\theta = 0$), more accurate results can be obtained than by Donnell's theory for a homogeneous shell. In this paper an attempt has been made to check the validity of this assumption for the case of an antisymmetrically laminated circular cylindrical shell. Solutions for the exact and modified theories are presented and the results are compared with those obtained from Donnell's approximation.

FORMULATION

Exact Theory

The governing differential equations of a laminated thin shell are:[5]

$$N'_x + \frac{1}{R} N^*_{x\theta} = 0 \tag{1}$$

$$N'_{x\theta} + \frac{N^*_\theta}{R} + \frac{Q_\theta}{R} = 0 \tag{2}$$

$$Q'_x + \frac{Q^*_\theta}{R} - \frac{N_\theta}{R} = \rho_1 \ddot{w} \tag{3}$$

$$M'_x + \frac{M^*_{x\theta}}{R} - Q_x = 0 \tag{4}$$

$$M'_{x\theta} + \frac{M^*_\theta}{R} - Q_\theta = 0 \tag{5}$$

Eliminating Q_x and Q_θ from eqn. (3), using (4) and (5), one can get

$$M''_x + \frac{2}{R} M'^*_{x\theta} + \frac{M^{**}_\theta}{R^2} - \frac{N_\theta}{R} = \rho_1 \ddot{w} \tag{6}$$

Equation (2) now takes the form

$$N'_{x\theta} + \frac{N^*_\theta}{R} + \frac{M'_{x\theta}}{R} + \frac{M^*_\theta}{R^2} = 0 \tag{7}$$

Thus, eqns. (1), (6) and (7) are the governing equations of motion of a laminated shell. The variation of forces and moments during vibration for an unsymmetrical cross-ply laminate are

$$
\begin{Bmatrix} N_x \\ N_\theta \\ N_{x\theta} \\ M_x \\ M_\theta \\ M_{x\theta} \end{Bmatrix} =
\begin{bmatrix}
A_{11} & A_{12} & 0 & B_{11} & B_{12} & 0 \\
A_{12} & A_{22} & 0 & B_{12} & \cdot B_{22} & 0 \\
0 & 0 & A_{66} & 0 & 0 & B_{66} \\
B_{11} & B_{12} & 0 & D_{11} & D_{12} & 0 \\
B_{12} & B_{22} & 0 & D_{12} & D_{22} & 0 \\
0 & 0 & B_{66} & 0 & 0 & D_{66}
\end{bmatrix}
\begin{Bmatrix} \varepsilon_x \\ \varepsilon_\theta \\ \varepsilon_{x\theta} \\ \kappa_x \\ \kappa_\theta \\ \kappa_{x\theta} \end{Bmatrix} \tag{8}
$$

in which the plate extensional, coupling and bending stiffnesses are given by

$$(A_{ij}, B_{ij}, D_{ij}) = \int_{-t/2}^{t/2} C_{ij}(1, Z, Z^2)\,dZ \qquad (9)$$

The reduced stiffnesses, C_{ij}, of an individual lamina are expressed in terms of the lamina principal material properties. Thus,

$$
\begin{aligned}
C_{11} &= E_1/(1 - v_{12}v_{21}) \\
C_{12} &= v_{12}E_2/(1 - v_{12}v_{21}) \\
C_{22} &= E_2/(1 - v_{12}v_{21}) \\
C_{66} &= G_{12}
\end{aligned}
\qquad (10)
$$

where $v_{21} = v_{12}E_2/E_1$.

The 1- and 2-directions coincide with the x- and θ-directions, respectively, of a cross-ply laminate. The variations of in-surface strains and changes of curvatures during vibration are:[6]

$$\varepsilon_x = u' \qquad \varepsilon_\theta = \frac{1}{R}(w + v^*) \qquad \varepsilon_{x\theta} = \frac{u^*}{R} + v'$$

$$\kappa_x = -w'' \qquad \kappa_\theta = \frac{1}{R^2}(v^* - w^{**}) \qquad \kappa_{x\theta} = \frac{1}{R}(v' - 2w'^*) \qquad (11)$$

The governing differential equations in terms of the displacements are obtained by substituting (11) in eqns. (1), (6) and (9) and using (8), (9) and (10). These are

$$A_{11}u'' + \frac{A_{12}}{R}(w' + v^*) - B_{11}w''' + \frac{B_{12}}{R^2}(v'^* - w'^{**})$$

$$+ \frac{A_{66}}{R}\left(\frac{1}{R}u^{**} + v'^*\right) + \frac{B_{66}}{R^2}(v'^* - 2w'^{**}) = 0$$

$$A_{66}\left(\frac{u'^*}{R} + v''\right) + \frac{D_{66}}{R^2}(v'' - 2w''^*) + \frac{A_{12}}{R}u'^* + \frac{1}{R^2}\left(A_{22} - \frac{B_{11}}{R}\right)$$

$$\times (w^* + v^{**}) - \frac{D_{12}}{R^2}w''^* + \frac{1}{R^3}\left(\frac{D_{22}}{R} - B_{11}\right)(v^{**} - w^{***}) = 0 \qquad (12)$$

and

$$B_{11}u''' + \frac{B_{12}}{R}(w'' + v''^*) - D_{11}w^{IV} + \frac{D_{12}}{R^2}(v''^* - w''^{**})$$

$$+ \frac{2}{R}B_{66}\left(\frac{u'^{**}}{R} + v''^*\right) + \frac{B_{12}}{R^2}u'^{**} + \frac{B_{22}}{R^3}(w^{**} + v^{***})$$

$$- \frac{D_{12}}{R^2}w''^{**} + \frac{D_{22}}{R^4}(v^{***} - w^{****}) + \frac{2}{R^2}D_{66}(v''^* - 2w''^{**})$$

$$- \frac{A_{12}}{R}u' - \frac{A_{22}}{R^2}(w + v^*) + \frac{B_{12}}{R}w'' - \frac{B_{22}}{R^3}(v^* - w^{**}) - \rho\ddot{w} = 0$$

Modified Theory $(\varepsilon_\theta = 0)$

Assuming that the cylinder is inextensible in the circumferential direction, i.e. $\varepsilon_\theta = 0$, the transverse displacements can be expressed as

$$w = -v^* \tag{13}$$

Accordingly, the variations of strains and changes in curvatures take the form

$$\varepsilon_x = u' \qquad \varepsilon_{x\theta} = v' + \frac{u^*}{R} \qquad \kappa_x = v''^*$$

$$\kappa_\theta = v^* + v^{***} \qquad \kappa_{x\theta} = v' + 2v'^{**} \tag{14}$$

The total energy of a layered shell having layers of equal thickness is

$$\phi = \iint (N_x\varepsilon_x + N_\theta\varepsilon_\theta + N_{x\theta}\varepsilon_{x\theta} + M_x\kappa_x + M_\theta\kappa_\theta + M_{x\theta}\kappa_{x\theta})\,d_x d_\theta$$

Substituting (14) and taking the variation of ϕ with respect to the virtual displacements δu and δv, one obtains after some algebra

$$N'_x + \frac{1}{R}N^*_{x\theta} = 0$$

$$M''_x + \frac{1}{R^2}(M^*_\theta + M^{***}_\theta) + \frac{1}{R}(M'_{x\theta} + 2M'^{**}_{x\theta}) + N'_{x\theta} - \rho\ddot{w}^* = 0 \tag{15}$$

These are same as in Ref. 4, since a layered shell with layers of equal thickness is considered. The governing equilibrium equations in terms of displacements in this case are given by

$$A_{11}u'' - B_{11}w''' - \frac{B_{12}}{R^2}(w' + w'^{**}) + \frac{A_{66}}{R}\left(\frac{u^{**}}{R} - w'\right)$$

$$- \frac{B_{66}}{R^2}(w' + 2w'^{**}) = 0$$

and

$$B_{11}u'''^* - D_{11}w^{IV*} - \frac{D_{12}}{R^2}(w''^* + w''^{***})$$

$$+ \frac{1}{R^2}\left[B_{12}u'^* - D_{12}w''^* - \frac{D_{22}}{R^2}(w^* + w^{***}) + B_{12}u'^{***} - D_{12}w''^{***}\right.$$

$$\left. - \frac{D_{22}}{R^2}(w^{***} + w^{*****})\right]$$

$$+ \frac{1}{R}\left[B_{66}\left(v'' + \frac{u'^*}{R}\right) + \frac{D_{66}}{R}(v'' - 2w''^*) + 2B_{66}\left(v''^{**} + \frac{u'^{***}}{R}\right)\right.$$

$$\left. + \frac{2D_{66}}{R}(v''^{**} - 2w''^{***})\right]$$

$$+ A_{66}\left(v'' + \frac{u'^*}{R}\right) + \frac{B_{66}}{R}(v'' - 2w''^*) - \rho\ddot{w}^* = 0 \qquad (16)$$

SOLUTION

For an antisymmetric cross-ply laminate, $A_{11} = A_{22}$, $B_{11} = -B_{22}$, $B_{12} = B_{66} = 0$ and $D_{11} = D_{22}$. The shell is assumed simply supported at both edges with boundary conditions

$$w = 0, \qquad M_x = 0, \qquad N_x = 0 \qquad \text{and} \qquad v = 0$$

These conditions and the differential eqns. (12) and (16) are satisfied by assuming the variations of u, v and w as

$$\begin{Bmatrix} u \\ v \\ w \end{Bmatrix} = \sum_{m=1}^{\infty}\sum_{n=1}^{\infty} \begin{Bmatrix} U_{mn} \cos\dfrac{m\pi x}{L} \cos n\theta \\ V_{mn} \sin\dfrac{m\pi x}{L} \sin n\theta \\ W_{mn} \sin\dfrac{m\pi x}{L} \cos n\theta \end{Bmatrix} \sin\omega t \qquad (17)$$

Substitution of (17) in (12) and in (16) yields a system of algebraic simultaneous equations in both the cases. For the case of exact solution, these are given by

$$a_1 U_{mn} + a_2 V_{mn} + a_3 W_{mn} = 0$$
$$a_2 U_{mn} + b_1 V_{mn} + b_2 W_{mn} = 0$$
$$a_3 U_{mn} + b_2 V_{mn} + (C_1 + \kappa_\omega^2 C_2) W_{mn} = 0 \qquad (18)$$

where

$$a_1 = -\left[\frac{C_{11}}{C_{66}} \left(1 + \frac{F}{2}\right) m^2\pi^2 \left(\frac{R}{L}\right)^2 + n^2 \right]$$

$$a_2 = m\pi n \frac{R}{L} \left(\frac{C_{12}}{C_{66}} + 1\right)$$

$$a_3 = m\pi \frac{R}{L} \left[\frac{C_{12}}{C_{66}} + \frac{C_{11}}{C_{66}} \left(\frac{F-1}{4\bar{n}}\right) \frac{m^2\pi^2}{z} \right]$$

$$b_1 = -m^2\pi^2 \left(\frac{R}{L}\right)^2 \left[1 + \frac{1}{12}\left(\frac{t}{R}\right)^2\right] - n^2 \frac{C_{11}}{C_{66}}$$

$$\times \left[\left(\frac{1+F}{2}\right) - \frac{t}{R}\left(\frac{F-1}{2\bar{n}}\right) + \left(\frac{1+F}{24}\right)\left(\frac{t}{R}\right)^2 \right]$$

$$b_2 = n \frac{C_{11}}{C_{66}} \left[(1+n^2)\left(\frac{F-1}{4\bar{n}}\right)\frac{t}{R} - \left(\frac{1+F}{2}\right) - n^2\left(\frac{1+F}{24}\right)\left(\frac{t}{R}\right)^2 \right]$$

$$- \frac{n}{6} m^2\pi^2 \left(\frac{t}{R}\right)^2 \left(\frac{R}{L}\right)^2 \left(1 + \frac{1}{2}\frac{C_{12}}{C_{66}}\right)$$

$$c_1 = \frac{C_{11}}{C_{66}} \left\{ -\left(\frac{1+F}{24}\right)\left(\frac{t}{R}\right)^2 \left[m^4\pi^4\left(\frac{R}{L}\right)^4 + n^4 \right] + \left(\frac{F-1}{2\bar{n}}\right)n^2\frac{t}{R} \right.$$

$$\left. - \left(\frac{1+F}{2}\right) \right\} - \frac{m^2n^2\pi^2}{6}\left(\frac{t}{R}\right)^2\left(\frac{R}{L}\right)^2 \left(\frac{C_{12}}{C_{66}} + 2 \cdot 0\right)$$

$$c_2 = \pi^4 \left(\frac{1+F}{24}\right)\left(\frac{R}{L}\right)^4 \left(\frac{t}{R}\right)^2 \frac{C_{11}}{C_{66}}$$

$$K_\omega^2 = \frac{\rho\omega^2 L^4}{D_{11}\pi^4}$$

$$z = \frac{L^2}{Rt}$$

and

$$F = \frac{E_2}{E_1}$$

In the case of modified theory two simultaneous algebraic equations are obtained. These are

$$d_1 U_{mn} + d_2 W_{mn} = 0$$

$$e_1 U_{mn} + (e_2 + n\kappa_\omega^2) W_{mn} = 0 \qquad (19)$$

where

$$d_1 = -\frac{C_{11}}{C_{66}}\left(\frac{1+F}{2}\right)m^2\pi^2\left(\frac{R}{L}\right)^2 - n^2$$

$$d_2 = \left[\frac{C_{11}}{C_{66}}\left(\frac{F-1}{4\bar{n}}\right)\frac{m^2\pi^2}{Z} - 1\right]m\pi\frac{R}{L}$$

$$e_1 = \frac{6mn}{\pi(1+F)}\frac{L}{t}\left[\frac{m^2(F-1)}{\bar{n}} - \frac{4Z}{\pi^2}\frac{C_{66}}{C_{11}}\right]$$

and

$$e_2 = n\left[m^4 + \left(\frac{L}{R}\right)^4\frac{(1-n^2)^2}{\pi^4}\right] + \frac{2m^2}{\pi^2(1+F)}\left(\frac{L}{R}\right)^2$$

$$\times \left\{\frac{C_{66}}{C_{11}}\left[(1-2n^2)^2 + 12\left(\frac{R}{t}\right)^2\right] - 2\frac{C_{12}}{C_{11}}n(1-n^2)\right\}$$

Equations (18) and (19) represent two systems of homogeneous algebraic simultaneous equations; the determinant of each of them should be zero for non-trivial solution. The characteristic equation of each of the determinants yields the frequencies of the shell in the respective cases.

RESULTS AND DISCUSSION

The results obtained by the above two methods, together with those using Donnell's approximation,[3] are shown in Tables 1 and 2 and Figs. 2–9. These include the results corresponding to orthotropic cases using the respective methods. The frequencies at various modal numbers using all the three methods are shown in Tables 1 and 2 and Fig. 2, whereas Figs. 3–9 represent the fundamental frequencies alone.

In the case of short shells (Table 1) it is observed that the frequencies corresponding to modified theory are much higher than the corresponding

TABLE 1

FREQUENCIES OF A LAYERED CYLINDRICAL SHELL
$(R/t = 100,\ Z = 10,\ E_1 = 30 \times 10^6\ \text{psi},\ E_2 = 0.75 \times 10^6\ \text{psi},\ G_{12} = 0.375 \times 10^6$ psi, $v_{12} = 0.25)$

Mode m n		Exact solution (number of layers)			$\varepsilon_\theta = 0$ theory (number of layers)			Donnell's theory (number of layers)		
		2	4	8	2	4	8	2	4	8
1	1	3·012	3·092	3·111	5·457	5·510	5·525	2·995	3·083	3·106
1	3	1·712	1·855	1·889	1·866	2·015	2·055	1·707	1·853	1·888
1	5	1·193	1·402	1·449	1·205	1·424	1·479	1·200	1·409	1·457
2	1	4·074	4·961	5·157	11·09	11·46	11·56	4·006	4·932	5·140
2	3	3·417	4·446	4·666	4·217	5·119	5·327	3·377	4·430	4·658
2	5	2·945	4·102	4·343	3·083	4·233	4·482	2·936	4·104	4.348
3	1	6·206	8·913	9·466	17·07	18·25	18·54	6·100	8·873	9·444
3	3	5·938	8·734	9·300	7·381	9·815	10·34	5·863	8·708	9·288
3	5	5·666	8·559	9·138	5·982	8·810	9·393	5·633	8·553	9·140
4	1	9·776	15·01	16·05	23·53	26·18	26·80	9·654	14·97	16·03
4	3	9·661	14·94	15·99	11·53	16·26	17·24	9·565	14·91	15·97
4	5	9·515	14·85	15·91	9·978	15·20	16·25	9·461	14·84	15·91
5	1	14·66	23·07	24·73	30·62	35·46	36·58	14·54	23·03	24·70
5	3	14·61	23·04	24·70	16·73	24·49	26·08	14·50	23·00	24·68
5	5	14·53	22·99	24·65	15·09	23·40	25·06	14·46	22·97	24·65

TABLE 2

FREQUENCIES OF A LAYERED CYLINDRICAL SHELL
$(R/t = 100,\ Z = 1000,\ E_1 = 30 \times 10^6\ \text{psi},\ E_2 = 0.75 \times 10^6\ \text{psi},\ G_{12} = 0.395 \times 10^6$ psi, $v_{12} = 0.25)$

Mode m n		Exact solution (number of layers)			$\varepsilon_\theta = 0$ theory (number of layers)			Donnell's theory (number of layers)		
		2	4	8	2	4	8	2	4	8
1	1	53·04	53·11	53·14	53·63	53·70	53·73	53·15	53·17	53·17
1	3	16·99	18·00	18·25	18·02	18·19	18·28	17·19	18·41	18·71
1	5	16·14	23·74	25·30	25·68	25·77	25·81	16·73	24·67	26·29
2	1	103·7	103·7	103·7	108·3	108·4	108·5	103·7	103·7	103·8
2	3	35·47	36·10	36·27	35·62	36·10	36·31	35·63	36·34	36·52
2	5	24·51	30·23	31·52	31·17	31·67	31·89	25·08	31·08	32·85
3	1	147·8	148·0	148·1	162·7	163·0	163·2	147·9	148·1	148·1
3	3	53·55	54·31	54·52	53·55	54·50	54·87	53·71	54·50	54·69
3	5	34·69	39·40	40·51	39·07	40·30	40·79	35·29	40·19	41·34
4	1	184·9	185·4	185·6	217·2	217·7	217·9	185·1	185·4	185·6
4	3	71·30	72·52	72·84	71·66	73·32	73·91	71·45	72·67	72·98
4	5	45·50	50·12	51·23	48·32	50·69	51·52	46·16	50·90	52·02
5	1	215·4	216·2	216·4	271·7	272·6	272·9	215·3	216·1	216·4
5	3	88·72	90·72	91·29	89·99	92·67	93·56	88·84	90·88	91·40
5	5	56·73	62·07	63·35	58·50	62·49	63·78	57·45	62·84	64·13

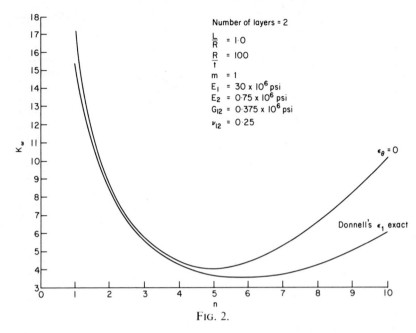

FIG. 2.

values from Donnell's approximation, particularly at lower modal numbers. However, this deviation from exact theory is less at very high frequencies. The frequencies of all modes of vibration with one nodal diameter are significantly affected due to this limitation. Further, the error in the frequencies is found to reduce as the number of layers increases. The frequencies of a long shell are presented in Table 2. The results obtained from modified theory compare well with exact theory only at lower frequencies, whereas results from Donnell's theory are in good agreement both at lower and higher frequencies. Again, the difference is quite considerable at $n = 1$, which decreases with the increase in the number of layers.

The variation of frequency with modal number n is shown in Fig. 2. The fundamental frequency for the shell occurs at modal number $m = 1, n = 5$ when modified theory is used, which changes to $m = 1, n = 6$ in the case of the other two theories. The frequencies determined by Donnell's theory coincide with the exact solution for all values of 'n' considered. The difference between the frequencies determined by modified and exact theory increases with 'n'.

The variations of the fundamental frequency with the geometrical non-dimensional parameters z and R/t and with E_1/E_2 are shown in Figs. 3–9.

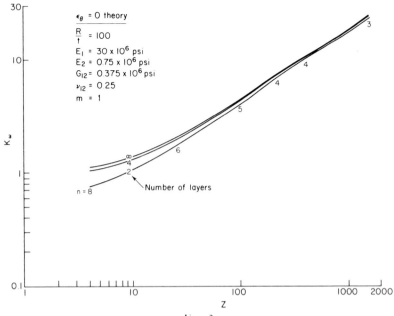

FIG. 3.

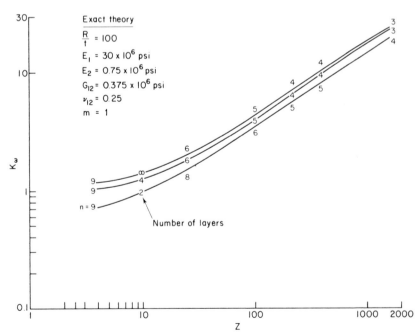

FIG. 4.

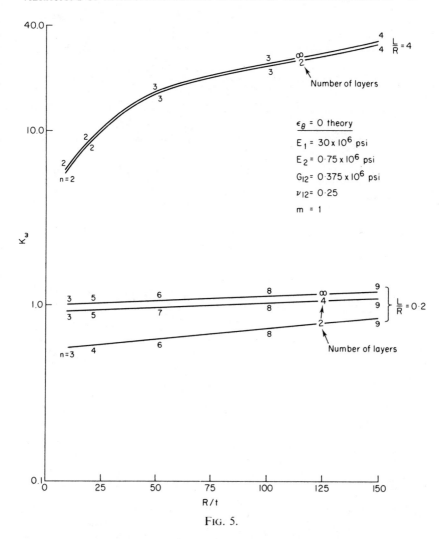

FIG. 5.

The effect of coupling between bending and extension is clearly brought out in all these figures. It is observed in all the cases that the orthotropic solution (i.e. $B_{11} = 0$) coincides with the solution of a shell with 8 layers. The variation of z using modified theory and exact theory are shown in Figs. 3 and 4. The modal number 'n' decreases as z increases. No change in modal number is observed at any z, even when the number of layers is increased, when modified theory is used (Fig. 3), whereas 'n' is found to decrease in the

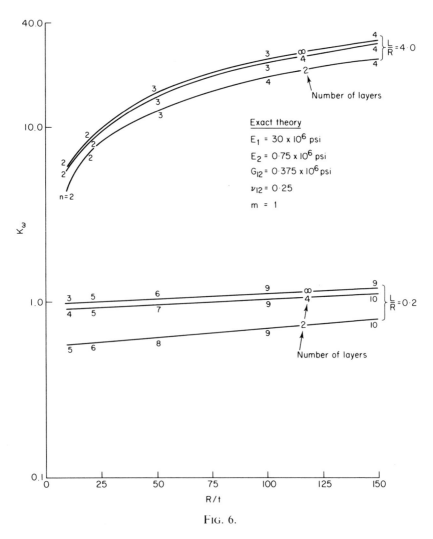

FIG. 6.

case of exact theory (Fig. 4). The variations of R/t for long and short shells using both theories are shown in Figs. 5 and 6. In the case of modified theory the effect of coupling is found negligible, at any particular R/t for long shells, whereas the same is significant for short shells (Fig. 5). This effect is predominant when exact theory is used (Fig. 6). In general, it can be observed that the effect of coupling dies down rapidly as the number of layers increases.

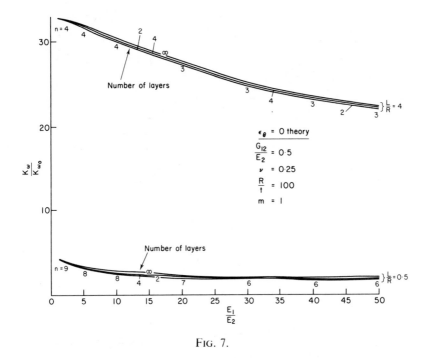

FIG. 7.

Frequencies have also been determined by using Donnell's theory, in addition to the above, for various values of R/t. The following observations have been made:

1. Low R/t: When the number of layers under consideration is small, modified theory results are closer to exact theory than those results obtained by using Donnell's. At higher $\bar{n}$, modified theory yields lower values and Donnell's higher values when compared to exact theory results. It has also been observed that for long shells, the results from modified theory compare better with exact theory than those from Donnell's.

2. High R/t: The results from Donnell's and modified theories are comparable to those from the exact theory, both yielding an upper bound. At low $\bar{n}$, results from Donnell's theory are closer to exact theory results but at high $\bar{n}$, values obtained from modified theory are closer.

The fundamental frequencies for various moduli ratios, E_1/E_2, are plotted in Figs. 7–9 after being normalised with the frequency at $E_1/E_2 = 1$.

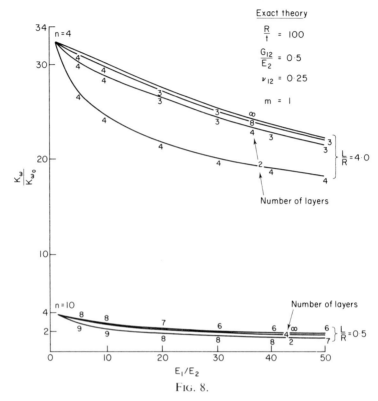

Fig. 8.

Once again, the coupling effect between stretching and bending is found insignificant in the cases of long and short shells (Fig. 7) when modified theory is used. This effect is predominant up to a value of $E_1/E_2 = 10$ for both long and short shells when exact theory is used (Fig. 8). For the case of a shell with two layers, Donnell's and modified theories yield frequencies more than those obtained by exact solution at any value of modulus ratio irrespective of whether the shell is long or short, as shown in Fig. 9. The results obtained by Donnell's approximation are closer to the exact results in both cases.

CONCLUSIONS

A theory assuming that the circumferential strain is zero is derived for an antisymmetrically laminated circular cylindrical shell together with the

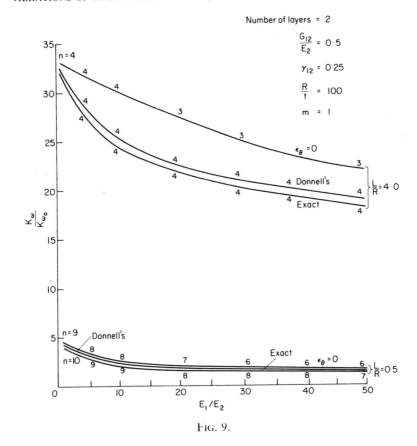

FIG. 9.

exact one. It is shown that the effect of coupling between bending and stretching cannot be clearly brought out when the former theory is used and that this yields comparable theory results for a shell with a large number of layers only.

REFERENCES

1. WHITNEY, J. M. and LEISSA, A. W. Analysis of heterogeneous anisotropic plates, *J. Appl. Mech.*, **36** (1969), pp. 261–6.
2. JONES, R. M. Buckling of unsymmetrically laminated cross-ply rectangular plates, *AIAA J.*, **11** (1973), pp. 1626–32.

3. JONES, R. M. and MORGAN, H. S. Buckling and Vibration of Cross-ply Laminated Circular Cylindrical Shells, AIAA 12th Aerospace Sci. Meeting, Washington, 1974.
4. YANG, Y. S., LEE, P. C. Y. and BILLINGTON, D. P. A simplified theory of thin cylindrical shells, *J. of Engng Mech. Div. ASCE*, EM4 (1974), pp. 719–36.
5. DONG, S. B. and TSO, F. K. W. On a laminated orthotropic shell theory including transverse shear deformation, *J. Appl. Mech.*, **39** (1972) pp. 1091–7.
6. KRAUS, H. *Thin elastic shells*, Wiley, New York, 1967.

Chapter 4

EFFECT OF SHEAR DEFORMATION AND ANISOTROPY ON THE NON-LINEAR RESPONSE OF COMPOSITE PLATES

A. K. Noor & S. J. Hartley

George Washington University Center at NASA–Langley Research Center, Hampton, Virginia, USA

SUMMARY

A study is made of the effects of variations in the geometry, lamination parameters and boundary conditions on the significance of transverse shear deformation and degree of anisotropy (non-orthotropy) of statically loaded composite plates. The analytical formulation is based on a form of the geometrically non-linear von-Karman type plate theory with the effects of transverse shear deformation, anisotropic material behaviour and bending–extensional coupling included. Numerical results are obtained by using displacement finite element models.

INTRODUCTION

Although several studies have been made on the effects of transverse shear deformation and material anisotropy (non-orthrotropy) on the response of composite plates, most of these studies are limited to linear problems (see, for example, Refs. 1–6). The present paper summarises the results of a recent study aimed at investigating the effects of variations in geometry, lamination parameters and boundary conditions on the significance of transverse shear flexibility and degree of anisotropy (non-orthotropy) on the non-linear response of statically loaded composite plates.

The analytical formulation is based on a form of the geometrically non-linear von-Karman type plate theory with the effects of transverse shear deformation, anisotropic material behaviour and bending–extensional coupling included.[7] Numerical studies are obtained using a higher order shear-flexible rectangular finite element model. The element has a total of 80 degrees of freedom and the shape functions used in approximating each of the displacement and rotation components consist of bicubic polynomials. Such a finite element model was shown to give highly accurate results for the response characteristics of the plate.[8]

NUMERICAL STUDIES

Numerical studies were conducted to investigate the effects of variations in the plate geometry, lamination parameters and boundary conditions on the non-linear response as well as on the significance of shear deformation and degree of anisotropy (non-orthotropy) in composite plates.

A quantitative measure for the relative importance of transverse shear deformation at different load levels is taken to be the ratio of the transverse shear strain energy to the total strain energy of the plate, U_{sh}/U. The measure for the degree of anisotropy (non-orthotropy) is taken to be the ratio of the contribution of anisotropic (non-orthotropic) terms to the total strain energy of the plate, U_{sh}/U. The two measures U_{sh}/U and U_a/U were first suggested and used in Ref. 5 for linear problems. They were also used in Ref. 9 to study the significance of shear deformation and degree of anisotropy on the post-buckling response.

The ratio U_{sh}/U can be thought of as the global error resulting from neglecting the transverse shear deformation, i.e. it is the error introduced by using the classical laminated plate theory in analysing the plate. Similarly, the ratio U_a/U represents the global error resulting from analysing an anisotropic plate as an orthotropic plate. Therefore, the two quantitative measures for transverse shear flexibility and degree of anisotropy used herein can provide a means for establishing the range of validity of the classical laminated and orthotropic plate theories. This is accomplished by examining the range of the different plate parameters and load levels for which U_{sh}/U and U_a/U are small (e.g. less than 5%).

Square plates having both symmetric and antisymmetric lamination with respect to the middle plane of the plate are considered. Also, four-layered and eight-layered quasi-isotropic plates are analysed. The plates are subjected to uniform transverse loading. The fibre orientations of the symmetric and antisymmetric laminates alternate between $+\theta$ and $-\theta$ with respect to the x_1-axis ($0 < \theta \le 45$). In the symmetrical laminates the $+\theta$

layers are at the outer surfaces of the laminate. The total thicknesses of the $+\theta$ and $-\theta$ layers in each laminate are the same. The four-layered and eight-layered quasi-isotropic laminates considered have fibre orientations of $+45/0/90/-45$ and $+45/0/90/-45/-45/90/0/+45$, respectively.

All numerical solutions presented herein are obtained by using a higher order, 16-node Lagrangian element and a 6×6 grid of elements in the full plate. Finer grids were used to ensure that adequate convergence was achieved by this grid. In all solutions, advantage was taken of the symmetries exhibited by the plate in the finite element analysis (see Ref. 9).

The material characteristics of the individual layers are taken to be those typical of high-modulus graphite–epoxy composites, namely

$$E_L/E_I = 40\cdot0 \qquad G_{LI}/E_I = 0\cdot6 \qquad G_{II}/E_I = 0\cdot5 \qquad v_{LI} = 0\cdot25$$

where subscript L refers to the direction of fibres and subscript T refers to the transverse direction, and v_{LI} is the major Poisson's ratio.

In addition to varying the boundary conditions, three parameters are varied, namely the number of layers NL, the fibre orientation angle θ and the thickness ratio of the plate h/a, where h and a are the thickness and side length of the plate. The number of layers NL was varied between one and

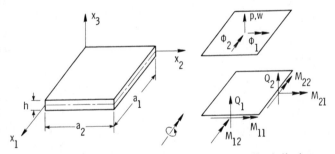

FIG. 1.   Sign convention for stress resultants and generalised displacements.

ten; θ was varied between $0°$ and $45°$; and h/a was varied between $0\cdot01$ and $0\cdot10$. Typical results showing the effects of variations in (a) the thickness ratio h/a, (b) the number of layers NL, (c) the fibre orientation angle θ and (d) the boundary conditions on the response of the plate, as well as on the significance of shear deformation and degree of anisotropy are presented in Figs. 2–6. Each figure is in four parts: (1) a load-deflection curve, specifically load versus transverse displacement at the centre, (2) a plot of the total strain energy as a function of the load, (3) a plot of the transverse shear strain energy U_{sh}/U and (4) a plot of the quantitative measure for the degree

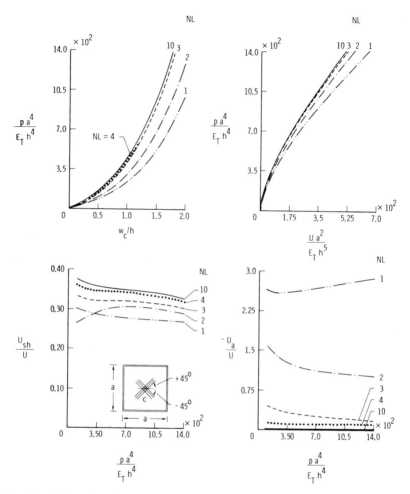

FIG. 2.   Effect of number of layers NL on the non-linear response of square plates with simply supported edges subjected to uniform transverse loading. $h/a = 0\cdot1$, fibre orientation $45/-45/\ldots$.

of anisotropy U_a/U as a function of the load. The results of the numerical studies can be summarised as follows:

1.   The stiffness of the plate sharply increases as the number of layers increases from 1 to 3, then becomes insensitive to further increase in the number of layers. The increase in the stiffness is associated with

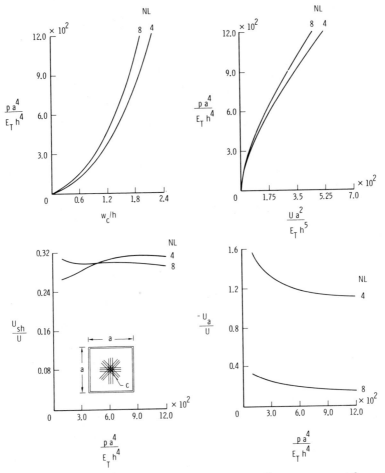

FIG. 3. Effect of number of layers NL on the non-linear response of quasi-isotropic square plates with simply supported edges subjected to uniform transverse loading. $h/a = 0\cdot1$, fibre orientation $45/0/90/-45/\ldots$.

    a decrease in the transverse displacement and the total strain energy (see Fig. 2).

2.   As the number of layers increases from 2 to 10, U_{sh}/U increases while U_a/U decreases. Generally both U_{sh}/U and U_a/U decrease with the increase in the load level. Exceptions to that are the increase in U_a/U for single-layered plates and the initial increase in U_{sh}/U for two-layered plates (see Fig. 2).

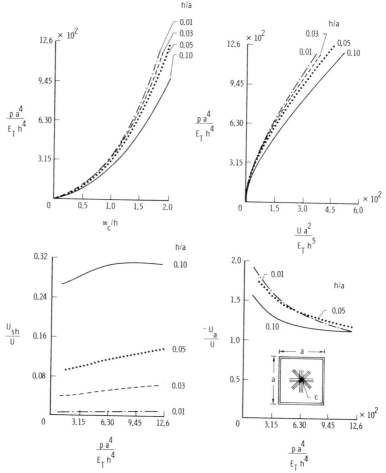

FIG. 4. Effect of thickness ratio h/a on the non-linear response of four-layered quasi-isotropic plate with simply supported edges subjected to uniform transverse loading. Fibre orientation $45/0/90/-45$.

3. The degree of anisotropy U_a/U is considerably higher in quasi-isotropic plates than in antisymmetrically laminated plates having the same number of layers and the same geometry (compare Figs. 2 and 3).

4. As expected, the stiffness increases with the increase in h/a. This is associated with an increase in U_{sh}/U and a decrease in U_a/U; and while U_{sh}/U for the medium thick and thin plates ($h/a \leq 0.05$)

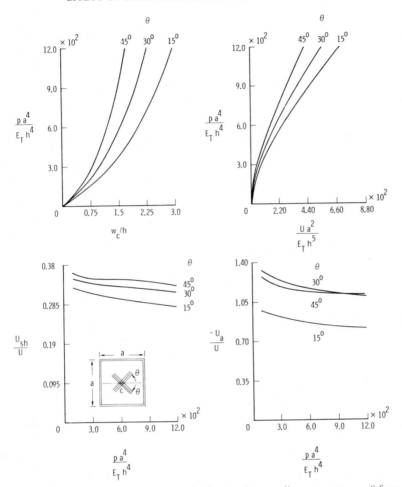

FIG. 5. Effect of fibre orientation angle θ on the non-linear response of four-layered antisymmetrically laminated plates with simply supported edges, subjected to uniform loading. $h/a = 0\cdot1$.

increases with the increase in the load level, U_a/U decreases with the increase in the load level (see Fig. 4).

5.   An indication of the relationship between the local errors resulting from neglecting the transverse shear deformation and the global error measure U_{sh}/U can be obtained from Fig. 4 by comparing the values of w/h, at different load levels, for plates with $h/a > 0\cdot01$ with those for the $h/a = 0\cdot01$ plate (which is essentially unaffected by

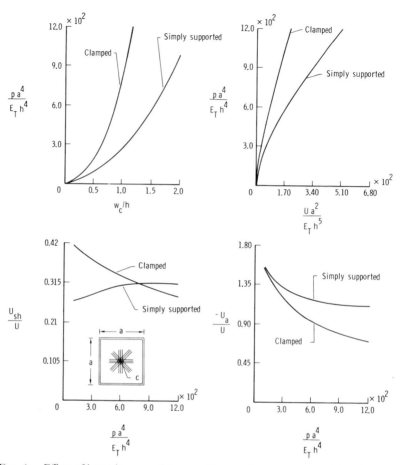

FIG. 6.   Effect of boundary conditions on the non-linear response of four-layered quasi-isotropic plate subjected to uniform transverse loading. $h/a = 0\cdot1$, fibre orientation $45/0/90/-45$.

transverse shear), and examining the corresponding values of U_{sh}/U for those plates. For example, at a load $p = 10^3 E_l h^4/a^4$, the difference between the maximum transverse displacement w/h for the thick and thin plates (with $h/a = 0\cdot1$ and $0\cdot01$) is over 20 %. The corresponding difference in U_{sh}/U is about 30 %.

6.   The stiffness of antisymmetrically laminated plates increases with the increase in the fibre orientation angle θ ($0 < \theta \leq 45$). The increase in the stiffness is associated with an increase in U_{sh}/U. Also,

U_a/U increases as θ increases from 15 to 30° then starts to decrease (see Fig. 5).

7. At low loads the shear deformation is much more pronounced in clamped plates than in simply supported plates having the same geometry and lamination. For four-layered quasi-isotropic plates, as the load increases, the shear deformation in clamped plates decreases while that in simply supported plates increases. For both clamped and simply supported plates U_a/U decreases with increasing load, the decrease being faster for clamped plates (see Fig. 6).

8. Since U_a is negative, all solutions obtained by neglecting the anisotropic terms (orthotropic plate theory) will overestimate the stiffness of the plate. Consequently, such analyses generally underestimate the maximum deflections of the plate.

On the basis of the numerical studies conducted, it appears that the classical plate theory is adequate for the non-linear analysis of uniformly loaded square plates with $h/a < 0.05$, and the orthotropic plate theory (with the anisotropic terms neglected) is adequate for the analysis of angle-ply plates with $NL > 4$. In general, if the effects of transverse shear deformation and/or anisotropy on the linear response are not significant, they will not be important in the non-linear analysis.

CONCLUDING REMARKS

Parametric studies were made of the effects of variations in geometry, lamination parameters and boundary conditions on the significance of transverse shear deformation and degree of anisotropy of statically loaded composite plates. The analytical formulation is based on a form of the geometrically non-linear von-Karman type plate theory with the effects of transverse shear deformation, anisotropic material behaviour and bending–extensional coupling included. The parametric studies show that for square composite plates subjected to uniform transverse loading, the classical plate theory is adequate for the non-linear analysis of plates with $h/a < 0.05$. In addition, the orthotropic plate theory is adequate for multilayered angle-ply plates with $NL \geq 4$. The results of the present study show that if the effects of transverse shear deformation and/or anisotropy on the linear response are not significant, they will not be important in the non-linear analysis.

REFERENCES

1. AMBARTSUMYAN, S. A., *Theory of Anisotropic Plates* (translated from Russian), Technomic, Stamford, Conn., 1970.
2. WHITNEY, J. M. and PAGANO, N. J., Shear deformation in heterogeneous anisotropic plates, *J. Appl. Mech.*, **37** (transactions of the ASME, 1970, p. 1031.
3. SRINIVAS, S. and RAO, A. K., Bending, vibration and buckling of simply supported thick orthotropic rectangular plates and laminates, *Int. J. Solids and Structures*, **6** (1970), p. 1463.
4. NOOR, A. K., Stability of multilayered composite plates, *Int. J. Fibre Science and Technology*, **8** (1975), p. 81.
5. NOOR, A. K. and MATHERS, M. D., Anisotropy and shear deformation in laminated composite plates, *AIAA J.*, **14** (1976), p. 282.
6. TURVEY, G. J. and WITTRICK, W. H., The large deflection and post-buckling behavior of some laminated plates, *Aeronautical Quarterly* (1973), p. 77.
7. NOOR, A. K. and MATHERS, M. D., Nonlinear finite element analysis of laminated composite shells, in *Computational Methods in Nonlinear Mechanics*, The Texas Institute for Computational Mechanics, Austin, Texas, 1974, p. 999.
8. NOOR, A. K. and MATHERS, M. D., Finite element analysis of anisotropic plates, *Int. J. Numerical Methods in Engineering* (in the press).
9. NOOR, A. K., MATHERS, M. D. and ANDERSON, M. S., Exploiting Symmetries for Efficient Post-buckling Analysis of Composite Plates, Proc. of AIAA/ASME/SAE 17th Structures, Structural Dynamics, and Materials Conf., King of Prussia, Penn., 1976, p. 39.
10. HEARMON, R. F. S., *An Introduction to Applied Anisotropic Elasticity*, Oxford University Press, 1961.

APPENDIX: FUNDAMENTAL EQUATIONS OF VON-KARMAN TYPE SHEAR-DEFORMATION PLATE THEORY

The fundamental equations of the von-Karman type shear-deformation plate theory used in the present study can be written in the following compact form (in index notation).

Equilibrium equations:

$$\partial_\alpha N_{\alpha\beta} = 0$$

$$\partial_\alpha Q_\alpha + N_{\alpha\beta} \partial_\alpha \partial_\beta w + p = 0$$

$$\partial_\alpha M_{\alpha\beta} - Q_\beta = 0$$

Constitutive relations:

$$N_{\alpha\beta} = C_{\alpha\beta\gamma\rho}(\partial_\gamma u_\rho + \tfrac{1}{2}\partial_\gamma w\,\partial_\rho w) + F_{\alpha\beta\gamma\rho}\,\partial_\gamma\phi_\rho$$

$$M_{\alpha\beta} = F_{\alpha\beta\gamma\rho}(\partial_\gamma u_\rho + \tfrac{1}{2}\partial_\gamma w\,\partial_\rho w) + D_{\alpha\beta\gamma\rho}\,\partial_\gamma\phi_\rho$$

$$Q_\alpha = C_{\alpha 3\beta 3}(\partial_\beta w + \phi_\beta)$$

In the above equations Greek indices take the values $1, 2$; $N_{\alpha\beta}$, Q_α and $M_{\alpha\beta}$ are the extensional, transverse shear and moment stress resultants; u^α, w and ϕ_α are the in-plane displacements, transverse displacement and rotation components of the middle plane of the plate; p is the intensity of the external transverse loading on the plate; $\partial_\alpha \equiv \partial/\partial x_\alpha$; and the $C_{\alpha\beta\gamma\rho}$, $D_{\alpha\beta\gamma\rho}$ and $F_{\alpha\beta\gamma\rho}$ are the extensional, bending and stiffness interaction coefficients of the plate, defined as follows:

$$(C_{\alpha\beta\gamma\rho}, F_{\alpha\beta\gamma\rho}, D_{\alpha\beta\gamma\rho}) = \sum_{k=1}^{NL} \int_{h_{k-1}}^{h_k} (1, x_3, x_3^2) c_{\alpha\beta\gamma\rho}^{(k)}\,dx_3$$

$$C_{\alpha 3\beta 3} = \sum_{k=1}^{NL} \int_{h_{k-1}}^{h_k} c_{\alpha 3\beta 3}^{(k)}\,dx_3$$

where $c_{\alpha\beta\gamma\rho}$ = plane stress reduced stiffness of the kth layer of the plate $= \bar{c}_{\alpha\beta\gamma\rho} - \bar{c}_{\alpha\beta 33}\bar{c}_{33\gamma\rho}/\bar{c}_{3333}$, with $\bar{c}$'s being the three-dimensional stiffness coefficients (Ref. 10).

Chapter 5

EXPERIMENTAL–NUMERICAL HYBRID TECHNIQUE FOR STRESS ANALYSIS OF ORTHOTROPIC COMPOSITES

K. Chandrashekhara & K. Abraham Jacob

Indian Institute of Science, Bangalore, India

SUMMARY

An experimental–numerical hybrid method is suggested for determination of stresses in two-dimensional orthotropic bodies. The method involves first rewriting the basic equations of orthotropic elasticity in terms of two stress parameters analogous to the sum and difference of normal stresses in isotropic case. These equations involve orthotropic constants and can be conveniently solved using a numerical procedure, which requires that boundary stresses are known. The boundary stresses are experimentally determined using techniques of photo-orthotropic elasticity. Two examples are given to illustrate the application of the method.

NOMENCLATURE

E_x, E_y Modulus of elasticity in x and y directions.
E_f, E_m Modulus of elasticity of fibre and matrix.
f Material fringe value.
G_{xy} Orthotropic shear modulus.
G_f, G_m Shear modulus of fibre and matrix, respectively.
h Finite difference mesh spacing.
i, j Integer variables denoting the position of mesh joint.
K_1, K_2 Orthotropic constants.
N Isochromatic fringe order.
x, y Cartesian co-ordinate axes.

∇^2 $\partial^2/\partial x^2 + \partial^2/\partial y^2$.

$\varepsilon_x, \varepsilon_y$ Strain components in x and y directions.

μ Poisson's ratio.

γ_{xy} Shear strain.

σ_x, σ_y Normal stress components.

τ_{xy} Shear stress component.

(Other symbols are explained in the text.)

INTRODUCTION

Techniques such as birefringent coatings, moiré, brittle lacquers, strain gauges and holography can be used on composites since they enable strains to be measured. However, a correct and accurate interpretation of results obtained by these methods is essential. The significance of this is demonstrated by Kedward and Hindle[1] by observing the effect of difference between the direction of principal stresses and strains in fibre-reinforced materials. Pih and Knight[2] initiated the use of transparent birefringent composites with anisotropic elastic and optical properties. Later Sampson[3] formulated a stress optic law which hypothesised the concept of Mohr's circle of birefringence. In addition Sampson introduced the concept that three photoelastic constants are required in order to characterise these new materials photoelastically. Dally and Prabhakaran[4, 5] suggested a method to predict the three fundamental photoelastic constants based upon properties of constituents. Bert[6] has shown that the concept of Mohr's circle of birefringence, as proposed by Sampson, is a direct result of tensorial nature of birefringence. Pipes and Rose[7] have shown that a single strain-optic coefficient coupled with four independent material properties are sufficient for prediction of the optical response of a birefringent anisotropic material. Recently, Prabhakaran[8] used a simplified form of stress-optic law proposed by Sampson to verify it for biaxial state of stress by employing strain gauge results on a undirectionally reinforced circular disk under diametral compression. Prabhakaran[9] has also determined the boundary stresses in an orthotropic ring subjected to diametral compression using the principles of photo-orthotropic elasticity.

However, a method of separating the stresses from photoelastic data obtained for a birefringent composite is yet to be developed. Sampson suggested that the integration of equilibrium equation can be used for the separation of stresses in photo-orthotropic elasticity and an application of this fact is reported by Knight.[10] This requires further study as there is still doubt about the meaning of isoclinics in photo-orthotropic elasticity.

In this paper an experimental–numerical hybrid technique is developed which simplifies the experimental stress analysis of orthotropic composites. In this case the stress need be obtained by experimentation only on the boundaries. The interior stresses can be obtained by a numerical procedure. The accuracy of the proposed hybrid technique is verified by examples.

BASIC EQUATIONS

The basic equations of elasticity as applied to two-dimensional orthotropic composites are:

(i) equilibrium equations,

$$\frac{\partial \sigma_x}{\partial x} + \frac{\partial \tau_{xy}}{\partial y} = 0 \qquad \frac{\partial \tau_{xy}}{\partial x} + \frac{\partial \sigma_y}{\partial y} = 0 \tag{1}$$

(ii) the compatibility equation,

$$\frac{\partial^2 \varepsilon_x}{\partial y^2} + \frac{\partial^2 \varepsilon_y}{\partial x^2} = \frac{\partial^2 \gamma_{xy}}{\partial x \, \partial y} \tag{2}$$

The stress–strain relations can be written as

$$\varepsilon_x = \frac{\sigma_x}{E_x} - \frac{\mu_x}{E_y}\sigma_y \qquad \varepsilon_y = \frac{\mu_y}{E_x}\sigma_x + \frac{1}{E_y}\sigma_y \qquad \gamma_{xy} = \frac{1}{G_{xy}}\tau_{xy} \tag{3}$$

Due to symmetry of stress–strain matrix

$$\mu_x E_x = \mu_y E_y \tag{4}$$

The compatibility eqn. (2) can now be written in terms of stresses as

$$\left(\frac{\partial^2}{\partial x^2} + K_1^2 \frac{\partial^2}{\partial y^2}\right)(\sigma_y + K_2^2 \sigma_x) = 0 \tag{5}$$

where

$$K_1^2 + K_2^2 = \frac{E_y}{G_{xy}} - 2\mu_x \tag{6}$$

$$K_1^2 K_2^2 = \frac{E_y}{E_x} \tag{7}$$

The two stress parameters S and D are defined such that

$$S = \sigma_y + K_2^2 \sigma_x \qquad D = \sigma_y - K_2^2 \sigma_x \tag{8}$$

i.e.

$$\sigma_y = \frac{S + D}{2} \quad \text{and} \quad \sigma_x = \frac{S - D}{2K_2^2} \tag{9}$$

Substituting relation (9) in (1) we get

$$\frac{\partial S}{\partial x} - \frac{\partial D}{\partial x} + 2K_2^2 \frac{\partial \tau_{xy}}{\partial_y} = 0 \qquad \frac{2\partial \tau_{xy}}{\partial x} + \frac{\partial S}{\partial y} + \frac{\partial D}{\partial y} = 0 \tag{10}$$

Differentiating first eqn. (10) with respect to x and second with respect to y and subtracting we get

$$\nabla^2 D = \frac{\partial^2 S}{\partial x^2} - \frac{\partial^2 S}{\partial y^2} - 2(1 - K_2^2) \frac{\partial^2 \tau_{xy}}{\partial x \partial y} \tag{11}$$

Differentiating, first, eqn. (10) with respect to y and second with respect to x and adding we get

$$\frac{\partial^2 \tau_{xy}}{\partial x^2} + K_2^2 \frac{\partial^2 \tau_{xy}}{\partial y^2} = -\frac{\partial^2 S}{\partial x \partial y} \tag{12}$$

Equation (5) can be written in terms of S as

$$\left(\frac{\partial^2}{\partial x^2} + K_1^2 \frac{\partial^2}{\partial y^2} \right) S = 0 \tag{13}$$

Equations (13), (12) and (11) can be solved in sequence to obtain the interior stress components from the known boundary values.

NUMERICAL PROCEDURE

For a two-dimensional orthotropic composite body subjected to loading, the boundary stresses can be determined at discrete points by any known experimental technique. From these boundary stresses, parameters S and D can be calculated using eqn. (8). Then eqn. (13) can be solved using the boundary values of S and the standard finite difference technique.[10] The recurrence relation for S at a point $0(i, j)$ (Fig. 1) can be written as

$$S_{i,j} = [S_{i+1,j} + S_{i-1,j} + (S_{i,j+1} + S_{i,j-1})K_1^2]/2(1 + K_1^2) \tag{14}$$

using relation (14) and, as iteration procedure, the values of S at the interior

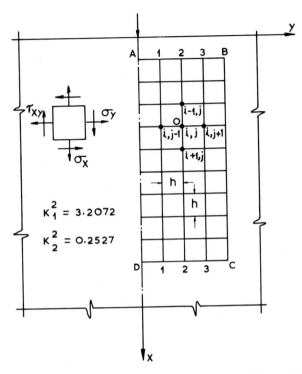

FIG. 1. A rectangular region (ABCD) in an orthotropic half-plane subjected to concentrated load.

mesh points can be calculated. Then, numerical value of the derivatives of S, namely $\partial^2 S/\partial x\,\partial y$, $\partial^2 S/\partial x^2$ and $\partial^2 S/\partial y^2$ at various mesh points can be calculated using finite difference. Then eqn. (12) can be solved to obtain the values of τ_{xy} at mesh points. The recurrence relation for eqn. (12) can be written as

$$(\tau_{xy})_{i,j} = (\tau_{xy})_{i+1,j} + (\tau_{xy})_{i-1,j} + K_2^2[(\tau_{xy})_{i,j+1} + (\tau_{xy})_{i,j-1}]$$

$$+ h^2 \frac{\partial^2 S}{\partial x\,\partial y} \bigg/ 2(1 + K_2^2) \qquad (15)$$

Knowing the boundary values of τ_{xy}, eqn. (15) can be used in an iteration programme to obtain the interior values of τ_{xy}. Then the numerical value of derivative of τ_{xy}, namely $\partial^2 \tau_{xy}/\partial x\,\partial y$ can be calculated using finite difference. As a third step eqn. (11) is solved in finite difference form to obtain the values of D at the interior mesh points. The recurrence relation for eqn. (11) becomes

$$D_{i,j} = \left[D_{i+1,j} + D_{i-1,j} + D_{i,j+1} + D_{i,j-1} - h^2 \left(\frac{\partial^2 S}{\partial x^2} \right)_{i,j} \right.$$

$$\left. + h^2 \left(\frac{\partial^2 S}{\partial y^2} \right)_{i,j} + 2h^2(1 - K_2^2) \left(\frac{\partial^2 \tau_{xy}}{\partial x \, \partial y} \right)_{i,j} \right] \Big/ 4 \qquad (16)$$

Now we have at interior mesh points the values of S, τ_{xy} and D. The values of stress components σ_x and σ_y can be calculated from eqns. (9).

APPLICATION OF PHOTO-ORTHOTROPIC ELASTICITY

The birefringence of orthotropic composites can be interpreted to give two relations, i.e.

$$\frac{\sigma_x}{f_x} - \frac{\sigma_y}{f_y} = N \cos(2\theta) \qquad (17)$$

and

$$\frac{\tau_{xy}}{f_{xy}} = \frac{N}{2} \sin(2\theta) \qquad (18)$$

where N is the optical (observed) isoclinic parameter. The value of N is obtained by considering the circle of birefringence[3] as

$$N^2 = (N_x - N_y)^2 + (2N_{xy})^2 \qquad (19)$$

where

$$N_x = \frac{\sigma_x}{f_x} \qquad N_y = \frac{\sigma_y}{f_y} \qquad \text{and} \qquad N_{xy} = \frac{\tau_{xy}}{f_{xy}}$$

The photoelastic data provide only the shear stress and a relation between the normal stresses. Along free boundaries, the boundary condition (component of stress normal to boundary is zero) allows determination of the components of stress separately. Along loaded boundaries, if the intensity of loading is known, then with the photoelastic data the complete information of the state of stress could be obtained.

Using the above procedure the boundary stress(es) can be determined experimentally, from which S may be computed along the boundaries. Then the values of S, τ_{xy} and D in the interior are determined using eqns. (13), (12) and (11), respectively, from which the individual values of normal stresses can be calculated. As the numerical procedure makes use of the experimentally determined boundary stresses for complete stress determination, the method can be referred to as the experimental–numerical hybrid technique.

EXAMPLES

First, the accuracy of the numerical method is verified by solving a problem for which a theoretical solution is available. A rectangular region ABCD (Fig. 1) is selected within a semi-infinite orthotropic plane subjected to concentrated load. The values of stresses σ_x, σ_y and τ_{xy} are computed along the boundaries of the region (for unit load) using the available theoretical solution.[13] The values of S and D along the boundaries are obtained from eqns. (8). The orthotropic constants used in this example are $K_1 = 3\cdot2072$ and $K_2 = 0\cdot2527$. The values of S, τ_{xy} and D are then computed in the interior using eqns. (14), (15) and (16), respectively. The results obtained from the numerical method at certain typical points are given in Table 1 and are compared with the theoretical results to indicate the convergence with the mesh size.

TABLE 1

THE VALUES OF S, D, τ_{xy}, σ_x AND σ_y AT CERTAIN POINTS, ILLUSTRATING THE EFFECT OF MESH SIZE

| X | Y | Mesh size | | | Exact | Quantity |
		$h = 0\cdot1$	$h = 0\cdot05$	$h = 0\cdot025$	(theoretical)	
0·2	0·1	0·946 125	0·951 177	0·950 686	0·950 540	
0·3	0·1	0·659 521	0·660 707	0·660 330	0·660 217	S
0·4	0·1	0·502 654	0·502 692	0·502 558	0·502 524	
0·2	0·1	−0·089 423	−0·009 659	−0·004 471	−0·005 106	
0·3	0·1	−0·295 681	−0·253 218	−0·254 490	−0·256 945	D
0·4	0·1	−0·319 954	−0·300 088	−0·301 938	−0·303 236	
0·2	0·1	0·904 66	0·945 435	0·945 293	0·945 435	
0·3	0·1	0·602 997	0·614 758	0·606 435	0·604 909	τ_{xy}
0·4	0·1	0·409 184	0·405 296	0·399 508	0·398 576	
0·2	0·1	2·048 988	1·901 141	1·889 903	1·890 871	
0·3	0·1	1·889 992	1·808 321	1·810 092	1·814 726	σ_x
0·4	0·1	1·623 680	1·588 406	1·591 802	1·594 303	
0·2	0·1	0·428 356	0·470 759	0·473 107	0·472 717	
0·3	0·1	0·181 920	0·203 745	0·202 920	0·201 636	σ_y
0·4	0·1	0·092 350	0·101 302	0·100 310	0·099 644	

As a second example, a square plate subjected to symmetric partial loading is considered (Fig. 2). An orthotropic birefringent model of the plate is made by casting epoxy resin (Araldite CY-230, 100 pbw + Hardener HY-951, 10 pbw—both manufactured by CIBA of India) around glass fibre (E-glass having same refractive index as the resin) unidirectionally arranged and stretched between two edges of a frame.

TABLE 2
PROPERTIES OF CONSTITUENT MATERIALS

Material	Material fringe constant kPa/m fringe (psi/in fringe)	Modulus of elasticity kPa (psi)	Poisson's ratio
Fibre (E-glass)	238 (1360)	7.24×10^{7} (10.5×10^{6})	0.2
Matrix (Araldite CY-230)	13.72 (79.5)	2.065×10^{6} (3.0×10^{5})	0.35

The composite specimens exhibited small amount of residual stress (0.167 of a fringe–uniform compression perpendicular to the fibres) due to shrinkage during curing of the resin. This was taken into account while determining the boundary stresses from the isochromatic pattern.

Bending specimens cut from the same material were used to determine the material fringe values f_x, f_y and f_{xy}. For the particular model used here, the following values were obtained:

$$f_x = 20.84 \ (119) \ \text{kPa/m fringe (psi/in fringe)}$$
$$f_y = 16.99 \ (97) \ \text{kPa/m fringe (psi/in fringe)}$$
$$f_{xy} = 16.11 \ (92) \ \text{kPa/m fringe (psi/in fringe)}$$

The material constants E_x, E_y, G_{xy} add μ_x are computed from the law of mixtures.[14]

$$E_x = V_f E_f + V_m E_m$$
$$E_y = E_f E_m / [V_f E_m + V_m E_f (1 - \mu_m^2)]$$
$$\mu_x = V_f \mu_f + V_m \mu_m$$
$$G_{xy} = G_f G_m / (V_m G_f + V_f G_m) \tag{20}$$

where f and m refer to fibre and matrix, respectively. The properties of the constituent materials are given in Table 2. The volume of fibres was computed from the weight of the fibre going into the composite (2 % in this

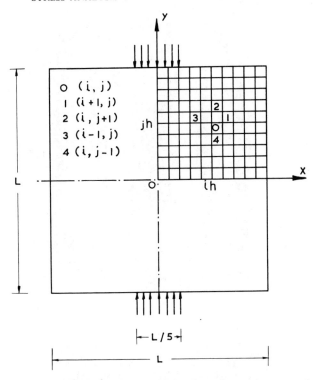

FIG. 2. Loading and finite difference mesh for the orthotropic square plate model.

case). The fibres were separated from the composite by burning off the resin at around 500 °C. The values of elastic constants obtained from computation were verified by testing the samples in tension and by measuring the strains. Using the above computed values of E_x, E_y, μ_x and G_{xy} the orthotropic constants K_1 and K_2 were determined using eqns. (6) and (7). The orthotropic constants for the material used in the present investigation are:

(i) fibres parallel to loading,

$$K_1^2 = 2{\cdot}05 \qquad K_2^2 = 0{\cdot}55$$

(ii) fibres perpendicular to loading,

$$K_1^2 = 1{\cdot}82 \qquad K_2^2 = 0{\cdot}487$$

The symmetric partial loading (concentration ratio 0·2) was applied

FIG. 3.   Light field isochromatic patterns for the square plate (fibres parallel to loading).

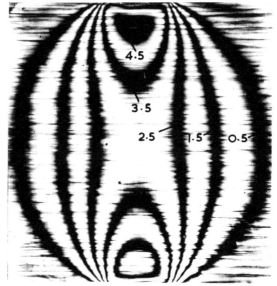

FIG. 4.   Light field isochromatic patterns for the square plate (fibres perpendicular to loading).

(Fig. 2) using a loading frame with lever loading arrangement. To obtain nearly uniform pressure distribution, an insert (2 mm rubber gasket sandwiched between 0·5 mm thick paper on either side) was used between the model and the loading pads.

Figures 3 and 4 show the dark and light field isochromatic patterns for the square plate for the cases of loading parallel and perpendicular to the direction of fibres.

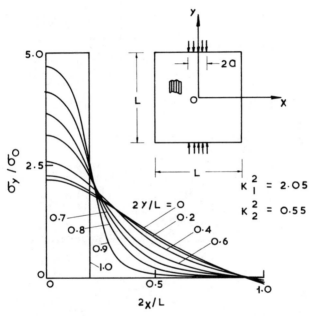

FIG. 5. Distribution of σ_x along different sections.

The values of S, τ_{xy} and D are obtained at the interior points using the numerical procedure described earlier. From this the distribution of stresses σ_x, σ_y and τ_{xy} along different sections of the composite plate are obtained and are presented in Figs. 5–10.

DISCUSSION AND CONCLUSION

From Table 1 it may be observed that the values of S can be determined at the interior points with reasonable accuracy by choosing an appropriate

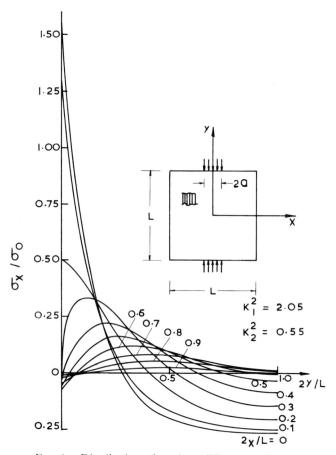

FIG. 6.   Distribution of σ_y along different sections.

mesh size. In this table a study of convergence of S, τ_{xy} and D has also been made. It may also be seen from the table that the stresses σ_x, σ_y and τ_{xy} obtained using the numerical procedure agree very well with exact theoretical results.

The second example illustrated the application of the numerical method to a practical problem. To verify the accuracy of the photoelastic measurement made while determining the boundary stress, the isochromatics was generated using the numerical method and this, along a particular section, has been compared with experimentally determined isochromatics in Fig. 11. It may be seen that there is a close agreement

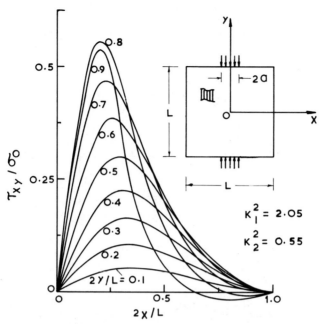

FIG. 7. Distribution of τ_{xy} along different sections.

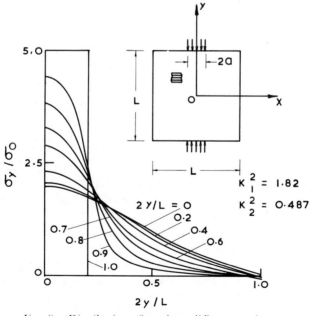

FIG. 8. Distribution of σ_x along different sections.

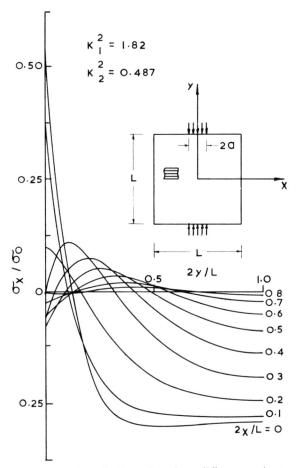

FIG. 9.   Distribution of σ_y along different sections.

between the two results, indicating that the boundary stresses for the numerical procedure have been accurately determined. Figure 12 gives a comparison of the difference in normal stresses obtained numerically with the theoretical results of Iyengar and Chandrashekhara.[15]

From the detailed study made here it may be concluded that the numerical method suggested can be effectively used for complete stress determination in orthotropic bodies. It may also be pointed out that the boundary stresses can be determined by using any one of the well-known experimental methods. For convenience's sake and illustration, photo-orthotropic elasticity was used here. With the availability of high-speed

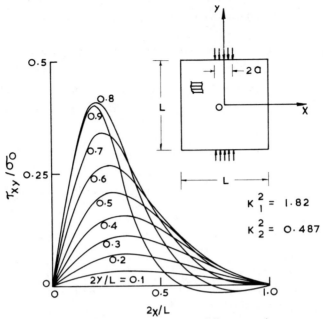

FIG. 10. Distribution of τ_{xy} along different sections.

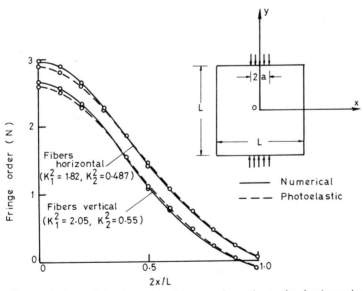

FIG. 11. Variation of isochromatic fringe orders along the horizontal axis obtained using the numerical procedure and comparison with the experimental values.

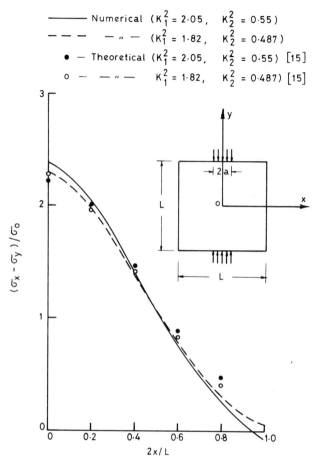

FIG. 12. Distribution of difference in normal stresses $(\sigma_x - \sigma_y)$ along the horizontal axis and comparison with the available theoretical results.

digital computers the whole numerical procedures can be programmed, thereby making the complete stress analysis simple.

REFERENCES

1. KEDWARD, K. T. and HINDLE, G. R. Analysis of strain in fibre reinforced materials, *J. Strain Analysis*, **5** (1970), No. 4, p. 309.

2. PIH, H. and KNIGHT, C. E. Photoelastic analysis of anisotropic fiber reinforced composites, *J. Composite Materials*, (1969), p. 94.
3. SAMPSON, R. C. A stress-optic law for photoelastic analysis of orthotropic composites, *Experimental Mechanics*, **10** (1970), No. 5, p. 210.
4. DALLY, J. W. and PRABHAKARAN, R. Photoelastic analysis of orthotropic composites, *Experimental Mechanics*, **11** (1971), No. 8, p. 346.
5. PRABHAKARAN, R. and DALLY, J. W. The application of photo-orthotropic elasticity, *J. Strain Analysis*, **7** (1972), No. 4, p. 253.
6. BERT, C. W. Theory of Photoelasticity of birefringent composites, 1st St. Louis Symp. on Advanced Composites (April 1971).
7. PIPES, R. B. and ROSE, J. L. Strain-optic law for certain class of birefringent composites, *Experimental Mechanics*, **14** (1972), No. 9, p. 355.
8. PRABHAKARAN, R. On the stress-optic law for orthotropic composite model materials in biaxial fields, *Experimental Mechanics*, **15** (1975), No. 1, p. 29.
9. PRABHAKARAN, R. Photoelastic analysis of a orthotropic ring under diametral compression, *AIAA Jl.*, **11** (1973), No. 6, p. 777.
10. KNIGHT, C. E., JR. Orthotropic photoelastic analysis of residual stresses in filament wound rings, *Experimental Mechanics*, **12** (1972), No. 2, p. 107.
11. ALLEN, D. V. DE G., *Relaxation methods*, McGraw-Hill, New York, 1954, p. 214.
12. DALLY, J. W. and RILEY, W. F. *Experimental stress analysis*, McGraw-Hill, New York, 1965.
13. LEKHNITSKII, S. G. *Anisotropic plates* (translated from the 2nd Russian edition by S. W. Tsai and T. Cheran), Gordon and Breach, 1968, p. 124.
14. SABODH, K. G., SVALBONAS, V. and GURTMAN, G. A. *Analysis of structural composite materials*, Marcel Dekker, New York, 1973.
15. SUNDARA RAJA IYENGAR, K. T. and CHANDRASHEKHARA, K. On the theory of the indentation test for the measurement of tensile strength of brittle materials, *Brit. J. Appl. Phys.*, **13** (Oct. 1962), pp. 501–7.

Chapter 6

PROPERTY OPTIMISATION ANALYSIS FOR MULTICOMPONENT (HYBRID) COMPOSITES

R. L. McCULLOUGH

University of Delaware, Newark, Delaware, USA

&

J. M. PETERSON

Boeing Commercial Airplane Company, Seattle, Washington, USA

SUMMARY

Multicomponent (hybrid) composite materials comprised of combinations of two or more families of fibres introduce additional degrees of compositional freedom for the construction of composite materials. This increased latitude provides additional means for tailoring composite materials by simultaneously adjusting several properties of the composite through the control of the concentration of the components. Systematic methods are presented for determining the concentration of components which can simultaneously optimise certain mechanical, thermal and electrical properties of a multicomponent composite. Particular emphasis is given to optimisations which minimise materials cost. It is shown that the optimisation procedure can be structured in the form of a classical linear programming problem. Specific criteria for material selection are presented for the case of a ternary composite comprised of two types of fibres and a binder phase.

INTRODUCTION

It has been well recognised that composite materials offer a unique advantage over traditional monolithic materials of construction in that their properties can be adjusted to efficiently match the mechanical, thermal and electrical requirements associated with specific applications. Early studies of composite materials were aimed at demonstrating the maximum achievable rigidity (or strength) of composite materials. Consequently, a single type of reinforcing fibre was usually selected to yield extreme performance characteristics. Performance optimisation centred on the selection of ply orientations to efficiently match attendant sets of properties to the anticipated mechanical loads (or deformations).

Recently, interest has turned to the use of hybrid systems comprised of two (or more) families of reinforcing fibres (and/or resins) in which properties, unique to each fibre family, can be combined to meet simultaneously two (or more) property requirements. The compositional latitude afforded by these multicomponent systems provides added flexibility in the design of composite material properties through the tailoring of the materials make-up of the composite. In particular, the additional degrees of freedom introduced by variations in composition point to the potential for achieving well-balanced performance character-istics by simultaneously adjusting several distinct properties of the multicomponent material system through control of the concentration of the various components.

The potential for tailoring composite systems to a balanced set of mechanical, thermal and/or electrical properties will gain added signific-ance as composite materials continue to evolve into structures intended to serve in multifunctional capacities. The use of composite materials in practical structures further demands a balance between the value of the improved performance and the materials cost. It can be anticipated that such performance optimisations will require that the characteristics of composite materials move from the extremes of a performance map toward more central, and lower cost, positions.

The purpose of this treatment is to develop systematic methods for analysing performance optimisations through the control of the materials make-up of multicomponent composite systems.

For these considerations, it will be sufficient to relate the properties of a typical unidirectional element (e.g. a ply) to the properties and concentration of the various components. The properties of this element can be related in subsequent treatments through well-defined transfor-mations to the properties of a structure composed of combinations of these elements in various orientations.

The basic role of the unidirectional element is well recognised. A large body of knowledge has been developed for predicting its properties from those of the components. The majority of this work, however, has focused on binary systems (one fibre, one resin). A simple extension of these models for estimating properties of multicomponent systems is discussed in the following section. It is shown that, for the purposes of this treatment, the constitutive relationships for several important properties can be cast into simple linear forms. Accordingly, initial material selection can be treated as a specialised Linear Programming problem. The basic notions underlying optimum material selection are illustrated by graphical techniques in the subsequent section. The graphical approach to the analysis is limited to ternary systems. An algebraic analysis is developed which can be used to treat performance optimisations for n-component systems. The results of the algebraic analysis for the special case of a ternary system (two fibre families and one binder phase) are used to develop specific criteria for materials selection. Particular emphasis is given to selections which minimise materials cost.

The optimisation methods are summarised in an outline format amenable to a computer analysis in the final section.

CONSTITUTIVE RELATIONSHIPS FOR ESTIMATING THE PROPERTIES OF MULTICOMPONENT COMPOSITES

The potential, as well as the limitations, of performance optimisations for multicomponent composites can be explored analytically by appropriate manipulations of constitutive relationships which relate the properties of the composite to the properties and concentrations of the components. Hence a prerequisite of a performance optimisation analysis is a collection of constitutive relationships which gives reasonable estimates of mechanical, thermal, and electrical properties in terms of the 'n' components of the composite system.

Considerable effort has been devoted to developing constitutive relationships to relate fibre and resin properties to the behaviour of typical unidirectional elements of binary composites. A variety of approaches has been used to develop constitutive relationships to describe the composite properties and concentrations. The results from these various studies have been summarised by several authors; among them Chamis and Sendeckyj,[1] Ashton et al.,[2] McCullough[3] and Jones.[4]

The results of these various approaches all reduce to the linear 'rule of mixtures' form for many of the important longitudinal properties. This form is consistent with a simple model which treats the response of the composite to mechanical, thermal or electrical loads by analogy to springs

TABLE 1

CONSTITUTIVE EQUATIONS FOR ESTIMATING LONGITUDINAL PROPERTIES OF TERNARY COMPOSITES

v = volume fraction; w = weight fraction

Properties	Symbol	Equation
Mechanical		
Modulus	E	$E = v_1 E_1 + v_2 E_2 + v_3 E_3$
Strength[a]	σ	$\sigma = v_1 \sigma_1 \lambda_1 + v_2 \sigma_2 \lambda_2 + v_3 \sigma_3 \lambda_3 \quad (\lambda_i = \varepsilon^*/\varepsilon_i)$
Poisson's Ratio	v	$v = v_1 v_1 + v_2 v_2 + v_3 v_3$
Thermal		
Coefficient of expansion	α	$\alpha = [v_1 E_1 \alpha_1 + v_2 E_2 \alpha_2 + v_3 E_3 \alpha_3]/[v_1 E_1 + v_2 E_2 + v_3 E_3]$
Thermal conductivity	K	$K = v_1 K_1 + v_2 K_2 + v_3 K_3$
Electrical		
Resistivity	ρ	$1/\rho = (v_1/\rho_1) + (v_2/\rho_2) + (v_3/\rho_3)$
Weight		
Density	d	$d = v_1 d_1 + v_2 d_2 + v_3 d_3$
Cost		
Cost/pound	C	$C = w_1 d_1 + w_2 d_2 + w_3 d_3$

[a] Calculated on the assumption that the composite fails at the lowest critical strain ε^* of the components; $\varepsilon^* = \min(\varepsilon_1, \varepsilon_2, \varepsilon_3)$.

and/or resistors connected in parallel. Under this simple model, factors such as the relative position of the components, packing geometry and fibre cross-sectional shape do not exert an influence on the composite properties. Accordingly, it is reasonable to expect the rule of mixtures form to yield adequate estimates for the longitudinal properties of multicomponent composites. Thence, many of the important properties may be estimated from equations of the form

$$P = \sum_{i=1}^{n} P_i v_i$$

where v_i is the volume fraction of the ith component and n is the total number of components. P is a particular property of the composite; P_i is the property of the ith component. The rule of mixtures estimate for some of the important mechanical, thermal and electrical properties for a three-component composite is summarised in Table 1. The constitutive relationship for longitudinal strength warrants further comment. The relationship given in Table 1 is based on the conservative notion that the strength of a collection of fibres is governed by the fibre component with the smallest elongation to break. This criterion is consistent with the traditional belief that materials with significant differences in breaking strains should not share the same load paths. Under this view, as the tensile load is increased, the collection of fibres is uniformly strained; eventually, a strain level is reached which corresponds to the smallest breaking strain of the fibre family within the collection. A subsequent infinitesimal increase in strain causes all those fibres characterised by the smallest breaking strain to fail. The sudden transfer of load to the remaining unbroken fibres is presumed to lead to catastrophic failure. Accordingly, from Hooke's law and the rule of mixture relationship for tensile modulus,

$$\sigma = \varepsilon E = \sum_{i=1}^{n} v_i E_i$$

where σ is the current level of tensile stress, ε is the current level of strain, E_i is the tensile modulus of the ith component, and v_i is the volume fraction of the ith component.

The strength of the ith component is given by

$$\sigma_i^* = \varepsilon_i^* E_i$$

where σ_i^* and ε_i^* are the ultimate stress and strain, respectively, for the ith (Hookeian) component.

According to this 'minimum strain criterion', the ultimate strength of the system is the stress level at which the elongation of the system has reached the ultimate elongation of the fibre family having the smallest strain to break, i.e., min $\{\varepsilon^*\}$. If ε_j^* is the smallest strain to break of the collection $\{\varepsilon_i^*\}$ so that $\varepsilon_j^* = \min \{\varepsilon_i^*\}$, then the ultimate strength of the system is given by:

$$\sigma^* = \varepsilon_j^* E = \varepsilon_j^* \sum_{i=1}^{n} v_i E_i$$

or:

$$\sigma^* = \sum_{i=1}^{n} v_i \lambda_i \sigma_i^*$$

where:

$$\lambda_i = \varepsilon_j^* / \varepsilon_i^*$$

with:

$$\sigma_i^* = E_i \varepsilon_i^*$$

and:

$$\lambda_i = \varepsilon_j^* / \varepsilon_i^* \quad \text{(note that } \lambda_j = 1)$$

By way of an example, typical values for the strain compatibility parameter, λ, are: glass, boron: $\lambda \doteq 0{\cdot}2$; glass, graphite: $\lambda \doteq 0{\cdot}1$, and graphite, boron: $\lambda \doteq 0{\cdot}9$. Hence, according to this strength criterion, combinations of carbon and boron fibres are strain compatible ($\lambda \sim 1{\cdot}0$) while combinations of (glass, boron) or (glass, graphite) are strain incompatible; i.e., $\lambda \sim 0{\cdot}1$.

Clearly, this particular criterion for strength drastically underestimates the potential strength for combined fibre systems since the model does not take cognizance of the ability of broken fibres to continue to contribute to the reinforcement via load transfer, through the binder phase, around a fibre break. However, in lieu of a more realistic estimate for strength, this relationship can be used to establish conservative estimates for longitudinal strength.

In addition to the property levels achieved, the effectiveness of a composite material will depend upon weight and cost. Thus, these characteristics must be included in the manifold of properties. The density and cost per pound of a composite are given as:

$$d = \sum_{i=1}^{n} d_i v_i$$

$$c = \sum_{i=1}^{n} C_i w_i$$

where d_i is the density of the ith component, C_i is the cost per pound of the ith component and w_i is the weight fraction of the ith component.

It is often useful to examine performance in terms of certain 'specific' properties (i.e. the property per unit weight) viz., $\bar{P} = P/d$. Substituting for P and d yields

$$\bar{P} = \left(\sum_{i=1}^{n} P_i v_i \right) \bigg/ \left(\sum_{k=1}^{n} d_k v_k \right)$$

This relationship may be rearranged to the form

$$\bar{P} = \sum_{i=1}^{n} \left[(d_i v_i) \bigg/ \left(\sum_{k=1}^{n} d_k v_k \right) \right] P_i/d_i$$

The term (P_i/d_i) is the specific property, $\bar{P}_i$, of the ith component. The term in brackets is the weight fraction, w_i, of the ith component:

$$w_i = (d_i v_i) \bigg/ \left(\sum_{k=1}^{n} d_k v_k \right)$$

Thence the specific property of a composite is related to the specific properties of the components through weight fractions rather than volume fractions as the composition variable; viz.,

$$\bar{P} = \sum_{i=1}^{n} \bar{P}_i w_i$$

The 'property' equations and the density equation have volume fractions as the composition variable while the 'specific property' equations and the cost equation have weight fraction as the composition variable. Either volume or weight fractions may be used as concentration variables in the analysis of performance; however, all relationships must be expressed in common variables. Volume fraction variables may be transformed to weight fraction variables by the expression

$$w_i = (d_i v_i) \bigg/ \left(\sum_{k=1}^{n} d_k v_k \right) = d_i v_i / d$$

Similarly, weight fraction variables may be transformed to volume fraction variables by:

$$v_i = \frac{w_i/d_i}{\sum\limits_{k=1}^{n} (w_k/d_k)} = d(w_i/d_i)$$

since:

$$\frac{1}{d} = \sum\limits_{k=1}^{n} (w_k/d_k)$$

The extension of the binary constitutive relationships for transverse properties to n-component systems is not as straightforward. In contrast to the longitudinal properties, transverse properties are sensitive to the relative locations of the reinforcing elements as well as to geometrical factors. However, it can be shown that a conservative (lower bound) estimate for transverse properties is obtained from a simple model which treats the composite response by analogy to springs and/or resistors connected in series. Again a linear form is obtained—but in terms of reciprocals of the various properties. Clearly, if the individual components are anisotropic (e.g., graphite fibres), then the inverse of the appropriate transverse properties must be used in the constitutive relationship. Since the lower bound model is insensitive to relative location, and geometrical factors, it can be readily extended to multicomponent systems; e.g.,

$$1/P = \sum\limits_{i=1}^{n} v_i/P_i$$

where P is a conservative estimate of a transverse property P, v_i is the volume fraction, and P_i is the corresponding transverse property of the ith component.

These constitutive relationships provide only crude estimates for the properties of multicomponent systems. However, a performance optimisation analysis is not necessarily concerned with the precise prediction of given properties. Indeed, the objective of such an analysis is not to predict the exact value of properties (such as modulus, cost, thermal conductivity, etc.), but rather to ensure that a collection of performance levels are met or, hopefully, exceeded. Consequently these conservative estimates for composite properties provide an adequate basis for performance optimisation analyses.

The additional latitude in achieving specified property levels for multicomponent composites can readily be illustrated by reference to a

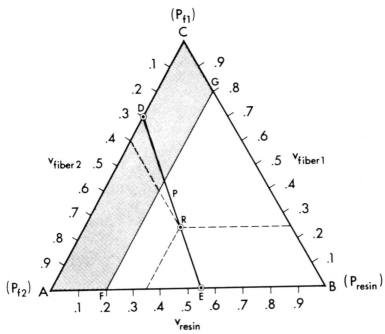

FIG. 1.   Typical format of ternary property maps.

ternary system. Because of material balance requirements (i.e., $v_1 + v_2 + v_3 = 1$), the performance maps for ternary systems can be displayed on familiar triangular diagrams.

Data presentation on a triangular diagram is based on the geometrical requirement that the sum of the distances, parallel to the three sides, from a point within an equilaterial triangle, is always equal to the length of a side of the triangle. Hence, by taking the lengths of each of the equivalent sides as unity and expressing the three components in fractional concentrations, it is possible to represent the composition of a ternary system by a single point within the diagram. A point situated on any one of the three sides indicates a binary system. A point located at the apices of the triangle represents the properties of a pure component.

Figure 1 illustrates the typical format for a ternary property map (e.g., two fibre families and one resin). The volume fraction of resin is given along the base line AB; the volume fractions of fibre family 1 and fibre family 2 are given along line BC and AC, respectively. The numbers in parentheses at the apices of the triangle represent the properties of the pure component corresponding to the composite property under consideration.

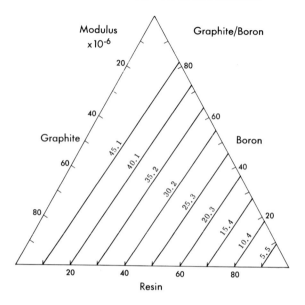

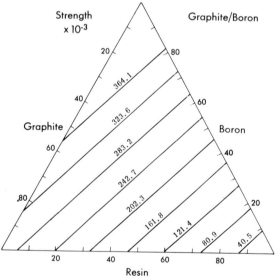

FIG. 2.   Selected property maps for the system graphite/boron/epoxy. The following typical values were used to generate the property maps: Young's modulus, in 10^6 psi (50/60/0·5); tensile strength, in 10^3 psi (300/445/5); density, in lbs/in^3 (0·060/0·098/0·040); critical strain, in per cent (0·6/0·7/10).

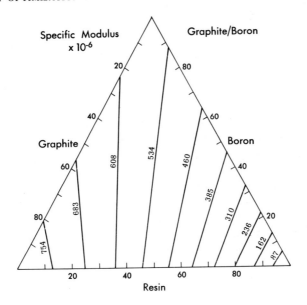

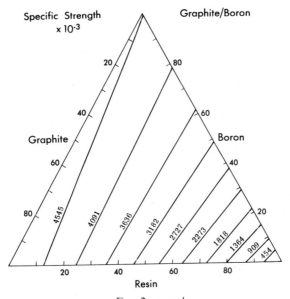

FIG. 2—*contd.*

The families of lines within the diagram specify arbitrarily selected property levels available to the ternary system. Thus, the line DE specifies the various ways of combining the three components to achieve a specified property level, P, of the composite. For example, the point R on line DE indicates that a 35% volume fraction of resin, a 25% volume fraction of fibre 1 (with property P_{j1}) and a 40% volume fraction of fibre 2 (with property P_{j2}) will result in a composite with a property P. Indeed any combination of resin and fibres 1 and 2 that falls on line DE will result in a composite with property P. For example, the point E on line DE offers the alternative concentration of 55% resin, 0% fibre 1 and 45% fibre 2 to achieve the same composite property P; while the point D offers yet another alternative of 0% resin, 70% fibre 1, and 30% fibre 2 to achieve property level P. Point D represents a bundle of fibres rather than a composite.

Some resin is required to ensure the integrity of the assembly of fibres. The smallest theoretical amount of resin (i.e., the most densely packed fibre assembly) that can be achieved is $\sim 10\%$ for ideal hexagonal packing of fibres or $\sim 20\%$ for ideal square packing of fibres. In practice volume fractions of resin greater than 30% are normally required. The area to the left of line FG in Fig. 1 represents all combinations of constituents that contain 20% or less of the binder phase; all combinations of constituents that fall within this area would be excluded for ideal square arrays. Hence the point which falls on the intersection of line DE and FG represents that combination of components with the minimum amount of resin (for an ideal square array) to yield a composite with property P; viz., 20% resin, 45% fibre 1, and 35% fibre 2.

Various levels for longitudinal modulus, longitudinal specific modulus, longitudinal strength and longitudinal specific strength for the ternary system graphite/boron/epoxy are given in Fig. 2 to illustrate the triangular property map display. Limiting resin lines ($\sim 30\%$) have been omitted for clarity.

The properties of a quaternary composite ($n = 4$) could be similarly presented as points within the volume of a tetrahedron. In such a display, the property level P would be defined by a plane rather than a line.

GRAPHICAL ILLUSTRATION OF PERFORMANCE OPTIMISATIONS

The important feature emphasised by the ternary property maps illustrated in the previous section is the additional latitude available in achieving a specified property level for a multicomponent composite system. This additional compositional freedom suggests that different properties could be simultaneously adjusted to satisfy a specified set of properties.

The property maps for two different composite properties, P and Q, are schematically superimposed in Fig. 3 to illustrate this point. For clarity, only one of the constant property levels of property P is shown.

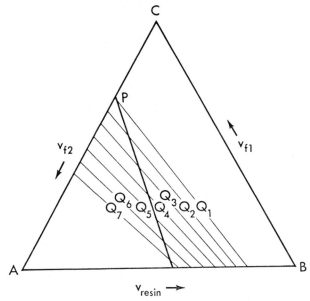

FIG. 3.   Superimposed property maps for properties 'P' and 'Q'.

Any combination of fibres and resin that falls on line P will yield a composite with property P. Similarly any combination of fibres and resin that falls on one of the lines, e.g., Q_j, representing a specified level of property Q will yield a composite with property Q_j. Any combination of components that simultaneously falls on lines P and Q_j will yield a composite with properties P and Q_j. Hence the intersection of lines P and Q_j uniquely determine the concentration of the three components required to yield a composite with properties P and Q_j.

Suppose for point of illustration that $Q_1 < \cdots Q_j < Q_{j+1} \cdots < Q_7$. If a composite is desired such that for a specified property P (e.g., modulus), the property Q (e.g., strength) is a maximum, then concentrations of components in the vicinity of the point determined by the intersection of P

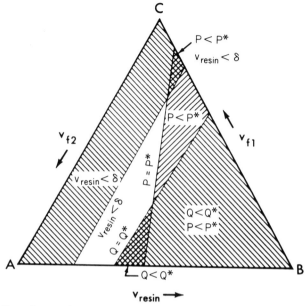

FIG. 4. Superimposed property maps illustrating bounding ranges on performance requirements.

and Q_7 should be considered. Alternatively, if a composite is desired with a specified property P (e.g., modulus) such that Q is minimum (e.g., cost per pound or thermal conductivity) then concentrations of components should be examined in the vicinity of the point determined by the intersection of P and a Q_j line consistent with the minimum feasible amount of resin. For example, if the resin limit line is taken as 30%, the intersection of P and Q_4 locates the region of interest for this hypothetical example.

In most cases of interest, property requirements will not specify that an exact property level P (e.g., no more or no less) be achieved. Usually a property level P or greater (or P or less for properties such as density and cost) will satisfy the performance requirements. Figure 4 illustrates superimposed maps for properties P and Q in which the various shaded regions represent concentration ranges excluded for failure to meet any one or all of the following requirements:

$$P > P^*$$

$$Q > Q^*$$

$$v_{\text{resin}} \geq \delta \quad \text{(the minimum feasible amount of resin)}$$

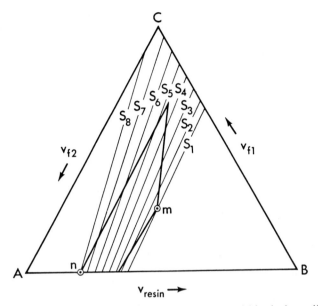

FIG. 5. Illustration of the optimisation of property 'S' within the bounding range
guaranteeing the performance requirements for properties 'P' and 'Q'.

A composite made up of components in the concentration ranges
represented by the unshaded area of the superimposed property maps will
meet or exceed the specified property requirements. Clearly, other limiting
property requirements could be superimposed to further restrict the
concentration range.

Of more importance, however, another property can be maximised (or
minimised) within the acceptable concentration range. Thus a material
composition could be specified such that property requirements $P^*, Q^*, \ldots,$
etc. are met or exceeded and the property S (e.g., cost per pound) is to be
minimal.

Figure 5 illustrates the bounding ranges of Fig. 4 superimposed on the
map for property S. Suppose that $S_1 < \cdots S_j < S_{j+1} \cdots < S_8$. The
concentration of components determined by point 'm' guarantees that
property requirement P^* and Q^* are met and that property S is minimal for
these requirements. Conversely, the concentration of components given by
point 'n' guarantees that the property requirements P^* and Q^* are exceeded
and that property S is maximal.

This schematic treatment illustrates the basic notions underlying a
performance optimisation analysis for multicomponent systems. Clearly, if

the number of components exceeds three, the graphical method cannot be used. In the following section algebraic relationships are developed to facilitate materials optimisations for n-component composite systems.

ALGEBRAIC ANALYSIS

For the purpose of an algebraic analysis, it will be convenient to abandon the 'P', 'Q' notation used in the previous section and adopt a compact index notation, viz., P_j = the jth distinct property of the composite (i.e., $P_1 = P$, $P_2 = Q$, $P_3 = S$, etc.); P_{ji} = the jth property of the ith component of the composite system.

The manifold of properties, summarised in Table 1, may be represented by equations of the general form:

$$P_j = \left(\sum_{i=1}^{n} P_{ji} u_{ji} x_i\right) \Big/ \left(\sum_{k=1}^{n} u_{jk} x_k\right) \tag{1}$$

where x may be either a volume fraction or weight fraction variable; and u_{jk} is a term that emerges from the transformation of concentration variables.

Let P_j^* be the lower limit that can be tolerated for property P_j such that the materials requirements may be expressed as

$$P_j^* \leq P_j$$

For these conditions eqn. (1) may be rearranged to:

$$0 \leq \sum_{i=1}^{n} u_{ji}[P_{ji} - P_j^*]x_i \tag{2}$$

Let $P_{j'}^{\dagger}$ be the upper limit that can be tolerated for property $P_{j'}$ such that the materials requirements may be expressed as

$$P_{j'}^{\dagger} \geq P_j$$

For these conditions eqn. (1) may be rearranged to:

$$0 \leq \sum_{i=1}^{n} u_{j'i}[P_{j'}^{\dagger} - P_{j'i}]x_i \tag{3}$$

Hence the materials requirements for a composite system, as represented by eqns. (2) and (3), can be expressed in a general form in terms of 'reduced' properties, ϕ_{ji}.

$$0 \leq \sum_{i=1}^{n} \phi_{ji} x_i \qquad (4)$$

where, depending upon the materials requirement (i.e., upper or lower bounds), the reduced property ϕ_{ji} is defined by eqns. (2) or (3).

The relationships between u_{ji} and ϕ_{ji} and the properties of the components are summarised in Table 2 for various properties formulated in terms of weight fractions and volume fractions and for the upper and lower bounds of the property requirements.

In addition to the constraints imposed by eqn. (4) to guarantee that the property requirements are met or exceeded, constraints to assure material balance and assembly integrity must also be formulated.

The material balance requirement is readily expressed as

$$\sum_{i=1}^{n} x_i = 1 \qquad \text{and} \qquad x_i \geq 0 \qquad \text{for all } `i` \qquad (5)$$

The integrity of the assembly may be ensured by requiring that a resin phase is always present in sufficient quantity to bind the reinforcing fibres together. Let δ be the minimum feasible volume fraction of resin required to bind the fibres together. Thus the integrity constraints may be written as:

$$0 \leq v_{\text{resin}} - \delta \qquad (6a)$$

for the volume fraction formulation and:

$$0 \leq (w_{\text{resin}}/d_{\text{resin}}) - \delta \sum_{i=1}^{n} (w_i/d_i) \qquad (6b)$$

for the weight fraction formulation. Equation (6b) points out a disadvantage of an analysis in the weight fraction formulation: the limiting weight fraction of resin varies with composition. Nonetheless in certain cases the simplicity of the property equations will outweigh this disadvantage.

Equations (4), (5) and (6) express the system of constraints that a composite material must satisfy in order to meet or exceed a prescribed set of properties.

TABLE 2

DEFINITIONS OF REDUCED PROPERTIES

$P_j^* =$ lower limit of property P_j; $P_j^\dagger =$ upper limit of property P_j; $\phi_{ji}^* = \mu_{ji}(P_{ji} - P_j^*)$; $\phi_{ji}^\dagger = \mu_{ji}(P_j^\dagger - P_{ji})$

Property		Formulation			
		Volume fraction (x = v)		Weight fraction (x = w)	
Description	Symbol P_j	μ_{ji}	ϕ_{ji}^*	μ_{ji}	ϕ_{ji}^*
Mechanical:					
Modulus	E	1	$E_i - E^*$	$1/d_i$	$(E_i - E^*)d_i$
Strength	σ	1	$\sigma_i \lambda_i - \sigma^*$	$1/d_i$	$(\sigma_i \lambda_i - \sigma^*)d_i$
Poisson's ratio	v	1	$v_i - v^*$	$1/d_i$	$(v_i - v^*)d_i$
Thermal:					
Coefficient of expansion	α	E_i	$E_i(\alpha_i - \alpha^*)$	$\bar{E}_i$	$\bar{E}_i(\alpha_i - \alpha^*)d_i$
Thermal conductivity	K	1	$K_i - K^*$	$1/d_i$	$(K_i - K^*)d_i$
Electrical:					
Resistivity	ρ	1	$[(1/\rho_i) - (1/\rho^*)]$	$1/d_i$	$[(1/\rho_i) - (1/\rho^*)]d_i$
Weight:					
Density	d	1	$d_i - d^*$	$1/d_i$	$(d_i - d^*)d_i$
Cost:					
Cost/Pound	C	d_i	$d_i(C_i - C^*)$	1	$C_i - C^*$
Specific:					
Specific modulus	$\bar{E}$	d_i	$d_i(\bar{E}_i - E^*)$	1	$\bar{E}_i - E^*$
Specific strength	$\bar{\sigma}$	d_i	$d_i(\bar{\sigma}_i \lambda_i - \bar{\sigma}^*)$	1	$\bar{\sigma}_i \lambda_i - \bar{\sigma}^*$

The expression for $\phi_{ji}^\dagger$ may be generated from the expression for ϕ_{ji}^* by replacing P_{ji}^* with $P_{ji}^\dagger$ and changing signs.

As illustrated in the previous section yet another property, H, can be maximised (or minimised) consistent with these constraints. Thus the analysis reduces to a constrained optimisation problem in which a collection of m different property requirements, $\{P_1, P_2, \ldots, P_m\}$, are to be met and the property H is to be maximised (or minimised). In summary, the performance optimisation analysis can be stated as:

Maximise (or minimise) H subject to the following $m + 2$ constraints:

'm' property constraints:

$$0 \leq \sum_{i=1}^{n} \phi_{1i} x_i$$

$$\vdots$$

$$0 \leq \sum_{i=1}^{n} \phi_{ji} x_i \tag{4}$$

$$\vdots$$

$$0 \leq \sum_{i=1}^{n} \phi_{mi} x_i$$

Material balance constraints:

$$1 = \sum_{i=1}^{n} x_i \tag{5}$$

Assembly integrity constraint:

$$0 \leq \begin{cases} v_{\text{resin}} - \delta \\ \qquad \text{or} \\ (w_{\text{resin}}/d_{\text{resin}}) - \delta \sum_{i=1}^{n} (w_i/d_i) \end{cases} \tag{6}$$

If the property (or characteristic) H can be expressed as a linear function of concentration, i.e.,

$$H = \sum_{i=1}^{n} H_i X_i$$

then the problem reduces to a classical Linear Programming problem for which well known techniques are available to obtain values for the concentration variables which simultaneously satisfy the m property requirements and maximise (or minimise) H. For example, several techniques for solving such Linear Programming Problems are discussed in detail by Loomba.[5]

The minimisation of material cost is an important example of a reduction to Linear Programming. In this case, the objective function is linear in the weight fraction variables; viz.,

$$H \to C = \sum_{i=1}^{n} C_i w_i$$

Accordingly, Linear Programming techniques can be used to determine the appropriate combination of components which satisfies the m property requirements and guarantee minimum materials cost.

For other cases, the objective function, H, may not be linear. In this event alternate techniques must be used to seek optimum values for the concentration variables. For example, if the objective function, H, has no extrema in the interior of the accessible region of the property map (e.g., the unshaded portion of Fig. 4), then the optimal solution will be found on the boundary of the accessible region. Furthermore, if the signs of the partial derivatives of H with respect to the concentration variables do not change within the accessible region, then any linear function, h, that increases (or decreases) in the same directions as H will have the same optimal solution as H (i.e., the same values of $x_1, x_2, \ldots, x_n$ that maximise or minimise h will also maximise or minimise H).

A linear function that increases or decreases in the same direction as H can be readily constructed from a Taylor series expansion of H. To avoid the redundancy introduced by the materials balance equation (5), the following substitution is made:

$$G(x_1, x_2, \ldots, x_n) = H(x_1, x_2, \ldots, x_{n-1}, 1 - x_1 - x_2, \ldots, - x_{n-1})$$

with:

$$g_0 \equiv G(0, 0, \ldots, 0)$$

$$g = g_0 + \sum_{i=1}^{n-1} h_i x_i$$

where:

$$h_i = \frac{\partial G}{\partial x_i}\bigg|_{x_1 = 0,\, x_2 = 0,\, \ldots,\, x_{n-1} = 0} \qquad (7)$$

so that

$$h = g - g_o = \sum_{i=1}^{n} h_i x_i \qquad (h_n \equiv 0) \qquad (8)$$

The function h is the required linear function that increases or decreases in the same direction as H.

Thus if H is monotonic in the accessible region of the property map, then the analysis may be conducted as a Linear Programming problem with the objective function taken as 'h' rather than 'H'. Subsequent substitution of the resulting values for $(x_1, x_2, \ldots, x_n)$ into the function H gives the true maximum or minimum values for H.

The constrained optimisation formulation provides a straightforward means of conducting a performance optimisation analysis. However, certain intermediate steps provide useful insights to material selection. Accordingly, it will be worthwhile to develop these intermediate analyses in order to obtain specific criteria for material selection.

Before proceeding with this development, it will be instructive to examine intuitively the constraining relationship for a simple case. For sake of discussion we will limit this examination to a ternary system, a volume fraction formulation, a single property P_j for which $u_{ji} = 1$ (for all i) and a lower bound specification so that $\phi_{ji} = P_{ji} - P_j^*$. We will further stipulate that component '1' is the resin phase and that components '2' and '3' are reinforcing fibres such that for property 'j', $P_{j3} > P_{j2} > P_{j1}$.

Equation (4) may be written out as

$$0 \leq \phi_{j1} v_1 + \phi_{j2} v_2 + \phi_{j3} v_3$$

Since the concentration variables are always positive there are only three conditions on the ϕ_{ji}s that will satisfy the materials requirement expressed by eqn. (4):

I $\phi_{j3} > \phi_{j2} > \phi_{j1} > 0$

II $\phi_{j3} > \phi_{j2} > 0 > \phi_{j1}$

III $\phi_{j3} > 0 > \phi_{j2} > \phi_{j1}$

The relationship between the materials requirement for the composite (i.e., $P_j \geq P_j^*$) and the properties of the components expressed by these three conditions for this simple example are schematically represented in Fig. 6.

Condition I represents a trivial situation in which the property requirements could be met or exceeded by any one of the components. However, if fibre components '2' or '3' are selected then some resin must be

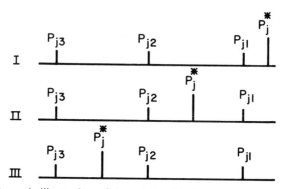

FIG. 6. Schematic illustration of the relationship between performance require-ment $(P_j \geq P_j^*)$ and the properties of the components of a ternary composite.

present to assure the integrity of the assembly of fibres, hence the maximum amount of component '2' or component '3' that would satisfy the integrity and material balance constraints is: max $v_2 = 1 - \delta$ or max $v_3 = 1 - \delta$. Then the range of volume fraction variables that could be used to satisfy the various constraints are:

Condition I

	max v_i	min v_i
v_1	1	δ
v_2	$1 - \delta$	0
v_3	$1 - \delta$	0

If no other property requirement was imposed on the system, then any combination of volume fractions in these ranges (that satisfy the material balance equation) would yield a composite that met or exceeded the property requirements. Since cost is, in many cases, proportional to the property level, the imposition of a material cost consideration would tend to drive the material system toward the pure resin phase. Conversely, the imposition of another property requirement might tend to drive the material system toward a more costly binary (or ternary) composite.

Condition II represents a situation in which the property requirements could be met or exceeded by either component '2' or '3' but not by component '1'; however, since some resin is required for integrity, at least a binary composite is required to meet the property requirements. The ranges are tabulated below.

Condition II

	max v_i	min v_i
v_1	$\phi_{j3}/(\phi_{j3} - \phi_{j1})$	δ
v_2	$1 - \delta$	0
v_3	$1 - \delta$	0

If no other property requirement is imposed on the system then a binary composite comprised of components '2' and '1' would satisfy the property requirements (provided P_{j2} is sufficiently greater than P_j^* to tolerate the dilution of properties due to the required presence of the resin phase). As in Condition I, cost considerations might support the binary systems comprised of components '1' and '2' while other property consideration might tend to drive the composition toward a binary composite of components '1' and '3' or a ternary composite.

Condition III represents a more interesting case than I or II. In this situation the property requirements can only be met or exceeded by component '3'; however, a ternary system could be constructed to satisfy the materials requirement. The possible concentration ranges are tabulated below.

Condition III

	max v_i	min v_i
v_1	$\phi_{j3}/(\phi_{j2} - \phi_{j1})$	δ
v_2	$[(1 - \delta)\phi_{j3} + \delta\phi_{j1}]/(\phi_{j3} - \phi_{j2})$	0
v_3	$1 - \delta$	$-[(1 - \delta)\phi_{j2} + \phi_{j1}]/(\phi_{j3} - \phi_{j2})$

Clearly a binary composite of components '3' and '1' would satisfy the property requirements. The ternary composite however, offers the advantage of requiring less of component '3' to offset the dilution of properties brought about by the presence of the resin phase in excess of δ.

TABLE 3
CONCENTRATION EXTREMES FOR CONDITIONS ON REDUCED PROPERTIES

'1' = resin phase; δ = minimum resin to bind fibres together; $F_{lmn} = \phi_{jl}/(\phi_{jm} - \phi_{jn})$

Condition	Maximum volume fraction			Minimum volume fraction		
	v_1	v_2	v_3	v_1	v_2	v_3
$\phi_{j3} > \phi_{j2} > \phi_{j1} > 0$	1	$1 - \delta$	$1 - \delta$	δ	0	0
$\phi_{j3} > \phi_{j2} > 0 > \phi_{j1}$	F_{331}	$1 - \delta$	$1 - \delta$	δ	0	0
$\phi_{j3} > 0 > \phi_{j2} > \phi_{j1}$	F_{331}	$(1 - \delta)F_{332} + \delta F_{132}$	$1 - \delta$	δ	0	0
$\phi_{j2} > \phi_{j3} > 0 > \phi_{j1}$	F_{221}	$1 - \delta$	$1 - \delta$	δ	0	0
$\phi_{j2} > 0 > \phi_{j3} > \phi_{j1}$	F_{221}	$1 - \delta$	$(1 - \delta)F_{223} + \delta F_{123}$	δ	0	$(1 - \delta)F_{223} + \delta F_{123}$
$\phi_{j3} > \phi_{j1} > 0 > \phi_{j2}$	1	$(1 - \delta)F_{332} + \delta F_{132}$	$1 - \delta$	δ	$(1 - \delta)F_{332} + \delta F_{132}$	0
$\phi_{j3} > 0 > \phi_{j1} > \phi_{j2}$	F_{331}	$(1 - \delta)F_{332} + \delta F_{132}$	$1 - \delta$	δ	0	F_{113}
$\phi_{j2} > \phi_{j1} > 0 > \phi_{j3}$	1	$1 - \delta$	$(1 - \delta)F_{223} + \delta F_{123}$	δ	F_{112}	0
$\phi_{j2} > 0 > \phi_{j1} > \phi_{j3}$	F_{221}	$1 - \delta$	$(1 - \delta)F_{223} + \delta F_{123}$	δ	0	0
$\phi_{j1} > \phi_{j2} > 0 > \phi_{j3}$	1	F_{112}	F_{113}	δ	0	0
$\phi_{j1} > 0 > \phi_{j2} > \phi_{j3}$	1	F_{112}	F_{113}	F_{221}	0	0
$\phi_{j1} > \phi_{j3} > 0 > \phi_{j2}$	1	F_{112}	$1 - \delta$	δ	0	0
$\phi_{j1} > 0 > \phi_{j3} > \phi_{j2}$	1	F_{112}	F_{113}	F_{331}	0	0

Thus for this condition the ternary system might offer a cheaper alternative to meeting or exceeding the property requirements.

The preceding analysis was based on the sequencing of the reduced properties ϕ_{ji} for an arbitrary indexing of the components. There is no guarantee (if a consistent indexing of components is followed) that the sequencing of the $\phi_{j'i}$s for one property requirement, P_j, will follow the same pattern as the sequencing of the $\phi_{j,i}$s for another property $P_{j'}$. Indeed in most cases sequencing will differ for the various properties, thus providing the motivation for material selection to optimise property trade-offs. In order to introduce a consistent indexing, we will arbitrarily index the components on densities to satisfy the inequality:

$$d_1 < d_2 < \cdots < d_n \tag{9}$$

Having thus fixed the indexing, there are

$$1 + (n - 1)n! \tag{10}$$

relevant conditions that will satisfy eqn. (4). The 13 conditions for a ternary system ($n = 3$) are given in Table 3 along with the maximum and minimum concentrations of the components that will satisfy the constraints. For simplicity, only the volume fraction formulation is given. Maxima and minima for the weight fraction formulation may be readily generated from transformation from volume fraction to weight fraction variables.

An intermediate analysis of interest is the selection of a material system in which a given property requirement is exactly met (no more or no less) while a second property is maximised or minimised. For example this problem would be encountered in the selection of a material system with the lowest materials cost that will achieve a specified property level (e.g., specific modulus).

For a ternary ($n = 3$) system, the property P_1 is achieved (no more or no less) under the conditions:

$$0 = \phi_{11}x_1 + \phi_{12}x_2 + \phi_{13}x_3$$

$$1 = x_1 + x_2 + x_3$$

It is convenient to define a dummy variable ξ (such that $x_1 = \xi$) and solve for the concentrations:

$$x_1 = \xi$$

$$x_2 = [\xi(\phi_{11} - \phi_{13}) + \phi_{13}]/(\phi_{13} - \phi_{12})$$

$$x_3 = [\xi(\phi_{12} - \phi_{11}) - \phi_{12}]/(\phi_{13} - \phi_{12})$$

TABLE 4

WEIGHT FRACTION OF COMPONENTS TO ACHIEVE SPECIFIC PROPERTY $\bar{P}$ AT MINIMUM COST

$$\phi_3 = \bar{P}_3 - \bar{P}; \quad \phi_2 = \bar{P}_2 - \bar{P}; \quad \phi_1 = \bar{P} - \bar{P}$$

Condition	's' Positive			's' Negative		
	w_1	w_2	w_3	w_1	w_2	w_3
$\phi_3 > \phi_2 > 0 > \phi_1$ $(\bar{P}_3 > \bar{P}_2 > \bar{P} > \bar{P}_1)$ (a)	α	$\dfrac{\phi_3(1-\alpha)+\alpha\phi_1}{\phi_3-\phi_2}$	$-\dfrac{(1-\alpha)\phi_2+\alpha\phi_1}{\phi_3-\phi_2}$	$\dfrac{\phi_3}{\phi_3-\phi_1}$	0	$\dfrac{-\phi_1}{\phi_3-\phi_1}$
(b)	$\dfrac{\phi_2}{\phi_2-\phi_1}$	$\dfrac{-\phi_1}{\phi_2-\phi_1}$	0			
$\phi_3 > 0 > \phi_2 > \phi_1$ $(\bar{P}_3 > \bar{P} > \bar{P}_2 > \bar{P}_1)$	α	$\dfrac{\phi_3(1-\alpha)+\alpha\phi_1}{\phi_3-\phi_2}$	$-\dfrac{(1-\alpha)\phi_2+\alpha\phi_1}{\phi_3-\phi_2}$	$\dfrac{\phi_3}{\phi_3-\phi_2}$	0	$\dfrac{-\phi_1}{\phi_3-\phi_1}$

δ = minimum volume fraction of resin required to bind fibres together; α = minimum weight fraction of resin required to bind fibres together; d = density; $s = [(\bar{P}_3 - \bar{P}_2)C_1 - (\bar{P}_3 - \bar{P}_1)C_2 + (\bar{P}_2 - \bar{P}_1)C_3](\bar{P}_3 - \bar{P}_2)$; $\alpha = \delta(d_1 d_3 \phi_3 - d_1 d_2 \phi_2)/\{d_2 d_3(\phi_3 - \phi_2) - \delta[d_2 d_3(\phi_3 - \phi_2) + d_1 d_3(\phi_1 - \phi_3) + d_1 d_2(\phi_2 - \phi_1)]\}$.

[a] $\alpha > \phi_2/(\phi_2 - \phi_1)$.
[b] $\alpha < \phi_2/(\phi_2 - \phi_1)$.

Substitution of these relations into the expression for another property P_2 yields a monotonic (but not necessarily linear) function in terms of ξ. Since P_2 is monotonic, the maximum (or minimum) will occur at the maximum or minimum values of ξ $(0 \le \xi \le 1)$. The appropriate choice of min ξ or max ξ for min P_2 or max P_2 will depend on whether P_2 is an increasing or decreasing function of ξ, i.e., whether the slope, $dP_2/d\xi$, is positive or negative. The sign of the slope is determined by the quantity

$$s = (\phi_{13} - \phi_{12})[\phi_{11}(P_{22}u_{23} - P_{23}u_{22})$$
$$+ \phi_{12}(P_{23}u_{21} - P_{21}u_{23}) + \phi_{13}(P_{21}u_{22} - P_{22}u_{21})]$$

Thus, depending on the sign of 's', the following choices would be made to minimise or maximise P_2

Choice of extremes:

For	$s > 0$	$s < 0$
max P_2	max ξ	min ξ
min P_2	min ξ	max ξ

Since the dummy variable was arbitrarily taken as $\xi = x_1$, the maximum or minimum values correspond to max x_1 and min x_1. These extremes for x_1 can readily be determined by establishing the proper sequence for $\phi_{11}, \phi_{12}, \phi_{13}$ and selecting the corresponding max x_1 or min x_1 from Table 3.

The problem can be further simplified for the special problem of selecting the lowest cost material system that will achieve a given level of some specific property, $\bar{P}$ (e.g., specific modulus or specific strength). For this special problem it is worthwhile to simplify the notation by the following definitions

$$P_1 = \bar{P} = \bar{P}_1 w_1 + \bar{P}_2 w_2 + \bar{P}_3 w_3$$
$$P_2 = C = C_1 w_1 + C_2 w_2 + C_3 w_3$$

with $\bar{P}_3 > \bar{P}_2 > \bar{P}_1$; $\bar{P}_1$ is taken as the specific property of the resin phase.

The appropriate weight fractions required to achieve a given value of $\bar{P}$ for the composite at minimum materials cost are summarised in Table 4. Notice that a negative cost slope, 's', always dictates a binary rather than a ternary system with the highest performance fibre (component '3') being the appropriate reinforcing fibre. A positive cost slope, s, usually implies a cost advantage for ternary systems except in those cases in which $\bar{P}_2$ is sufficiently greater than $\bar{P}$ to tolerate the dilution of the properties due to the presence of the required resin phase.

TABLE 5

EXTREME LIMITS ON THE CONCENTRATION OF THE COMPONENTS OF A TERNARY COMPOSITE WHICH SATISFY PROPERTY CONSTRAINTS

$$\phi_{11} > \phi_{12} > 0 > \phi_{13}$$
$$\phi_{21} > 0 > \phi_{22} > \phi_{23}$$
$$d_1 < d_2 < d_3$$

Minimum limiting concentrations

	Component -1	Component -2	Component -3
P_1: $(\phi_{11} > \phi_{12} > 0 > \phi_{15})$	$x^*_{11} = \delta$	$x^*_{12} = 0$	$x^*_{13} = 0$
P_2: $(\phi_{21} > 0 > \phi_{22} > \phi_{23})$	$x^*_{21} = \delta$	$x^*_{22} = 0$	$x^*_{23} = -\dfrac{[(1-\delta)\phi_{22} + \delta\phi_{21}]}{(\phi_{23} - \phi_{22})}$

Maximum limiting concentration

	Component 1	Component 2	Component 3
P_1: $(\phi_{21} > \phi_{12} > 0 > \phi_{13})$	$x^\dagger_{11} = \phi_{13}/(\phi_{13} - \phi_{11})$	$x^\dagger_{12} = 1 - \delta$	$x^\dagger_{13} = 1 - \delta$
P_2: $(\phi_{21} > 0 > \phi_{22} > \phi_{23})$	$x^\dagger_{21} = \phi_{23}/(\phi_{22} - \phi_{21})$	$x^\dagger_{22} = \dfrac{[(1-\delta)\phi_{23} + \delta\phi_{21}]}{(\phi_{21} - \phi_{22})}$	$x^\dagger_{23} = 1 - \delta$

A second useful intermediate analysis is the determination of whether or not the current material selection can indeed satisfy all the 'm' property requirements. Again, this analysis can be illustrated by reference to a ternary system ($n = 3$) for which two distinct property levels are desired ($m = 2$). For this case the constraining equations may be written out as:

$$0 \leq \phi_{11}x_1 + \phi_{12}x_2 + \phi_{13}x_3$$

$$0 \leq \phi_{21}x_1 + \phi_{22}x_2 + \phi_{23}x_3$$

$$1 = x_1 + x_2 + x_3$$

$$0 \leq v_1 - \delta$$

component '1' is taken as the resin phase based on the indexing of densities (i.e. $d_1 \leq d_2 \leq d_3$). For the purpose of illustration we will further stipulate

$$\phi_{11} > \phi_{12} > 0 > \phi_{13}$$

$$\phi_{21} > 0 > \phi_{22} > \phi_{23}$$

The limiting concentrations can be readily obtained from Table 3. For compactness in notation, it will be convenient to label the max (x_i) corresponding to property 'j' as $x^\dagger_{ji}$ and the min (x_i) corresponding to property 'j' as x^*_{ji}. The resulting maximum and minimum limiting concentrations corresponding to the property requirements P_1 and P_2 are summarised in Table 5.

The smallest value of $x^\dagger_{ji}$, i.e., min ($x^\dagger_{1i}, x^\dagger_{2i}, \ldots$, or x^*_{ni}) represents the maximum amount of the ith component that will yield a composite with properties that meet or exceed property levels P_1 and P_2. The largest value of x^*_{ji}, i.e., max ($x^*_{1i}, x^*_{2i}, \ldots$, or x^*_{ni}), the minimum amount of the ith component that can be used in order to meet or exceed P_1 and P_2. Thus the concentration of the ith component must fall within x^*_i [$= \max(x^*_{ji})$] $\leq x_i \leq x^\dagger_i [= \min(x^\dagger_{ji})]$ in order to satisfy the property requirements.

For the current example these concentration bounds may be written out as:

$$\max(x^*_{11} \text{ or } x^*_{21}) \leq x_1 \leq \min(x^\dagger_{11} \text{ or } x^\dagger_{21})$$

$$\text{or } \delta \leq x_1 \leq \min(x^\dagger_{11} \text{ or } x^\dagger_{21})$$

$$\max(x^*_{12} \text{ or } x^*_{22}) \leq x_2 \leq \min(x^\dagger_{12} \text{ or } x^\dagger_{22})$$

$$\text{or } 0 \leq x_2 \leq \min(x^\dagger_{12} \text{ or } x^\dagger_{22})$$

$$\max(x^*_{13} \text{ or } x^*_{23}) \leq x_3 \leq \min(x^\dagger_{13} \text{ or } x^\dagger_{23})$$

$$\text{or } \max(x^*_{13} \text{ or } x^*_{23}) \leq x_3 \leq 1 - \delta$$

Any concentration of the components within these ranges that satisfy the material balance $(x_1 + x_2 + x_3 = 1)$ will yield a composite whose properties meet or exceed the requirements for properties P_1 and P_2.

Clearly, any 'solution' which results in x_i^* [$= \max (x_{ji}^*)] > x_i^\dagger$ [$= \min (x_{ji}^\dagger)$] indicates that not all 'm' property requirements can be simultaneously met or exceeded. If $x_i^* = x_i^\dagger$ for all 'i', no further optimisation of a property H will be possible. The value of H for $x_i = x_i^* = x_i^\dagger$ is optimal.

CONCLUSIONS

The preceding treatment illustrates a systematic method for analysing the performance of multicomponent composite systems. The key element of the treatment was the structuring of the problem into a constrained optimisation algorithm; cases which reduce to classical Linear Programming problems are illustrated. The reduction to linear programming provides a basic and readily available computer technique to conduct the analysis.

It will be useful to summarise the analysis in outline form.

Procedure Outline
1. Input Requirements
   A. Mathematical models:
   Linear expressions which adequately estimate the properties of composites in terms of the properties and concentrations of components (e.g., Table 1).
   B. Materials data:
   P_{ji}: Values for the jth property of the ith component corresponding to the jth property of the composite.
   δ: The minimum amount of a resin phase that is required to ensure the integrity of the assembly.
   C. Design Data:
   P_j^* (and/or $P_j^\dagger$): The lower (or upper) limits that may be tolerated for the jth property of the composite.
   H (or h): An algebraic expression (with appropriate numerical values of the parameters) which relates the property (or characteristic) to be optimised to the concentration of the components (e.g., cost of materials).

II. Initial Constructions and Test

A. Form reduced properties, ϕ_{ji}, similar to Table 3.

Test: If for any 'j', ϕ_{ji} is negative for all 'i', reject current material system.

B. Construct system of:

'm'-property constraints (eqn. 4).

Material balance constraints (eqn. 5).

Composite integrity constraint (eqn. 6).

III. Determine Feasibility of Current Materials Selection

(If $n > 3$, it will be more efficient to by-pass this step and accept possible rejection of the material system in step IV as an unfeasible solution).

A. Determine limits on concentration of components:

Construct max (x_i), min (x_i) array similar to Table 3 ($2n$ columns; 1 + $(n - 1)n!$ rows).

Establish the sequencing of the current ϕ_{ji}s with respect to 'i' for each 'j'.

Select max (x_i), min (x_i) values from array corresponding to the resulting sequence of the current ϕ_{ji}s.

Assign max (min x_i) as the lower limit (x_i^*), and min (max x_i) as the upper limit $(x_i^{\dagger})$ of each x_i.

B. Test limits:

If $x_i^* < 0$, $x_i^{\dagger} < 0$, or $x_i^{\dagger} < x_i^*$ for any 'i', reject current material system. If $x_i^* = x_i^{\dagger}$ for all 'i' system is determined, no further optimisation of H will be possible. The value for $H(x_i = x_i^*, x_2 = x_2^*, \ldots, x_n = x_n^*)$ is optimal.

IV. Determine Optimum Materials Design for Property (or Characteristic) H—Linear Programming Case

A. Set up constraints (II.B).

B. For H a linear function:

Assign H as objective function.

Maximise (or minimise) H as dictated by the problem.

Tabulate resultant values of $(x_1, x_2, \ldots, x_n)$ as the optimum materials composition for the current materials system.

Tabulate resultant maximum (or minimum) value of H as optimum for current input.

C. For H a 'monotonic' function in the accessible region:

Construct h_is from H (eqn. 7).

Assign h (eqn. 8) as the objective function.

Maximise (or minimise) h as dictated by the requirements of the problem for a maximum (or minimum) for H.

Tabulate resultant values of $(x_i, x_i, \ldots, x_n)$ as the optimum materials composition for current input.

Substitute resultant values of $(x_1, x_2, \ldots, x_n)$ into H and tabulate resultant value as optimum for the current input.

D. Determine Properties of Optimum System:

Substitute resultant values of $(x_1, x_2, \ldots, x_n)$ into appropriate property equations and tabulate resultant P_js as the properties of the optimum system.

The present treatment focused on a simple uniaxial structural element. The results can therefore be applied directly to simple structures (e.g., struts and simple beams). Clearly, more complex structural elements will require a more involved treatment. Nonetheless, the general format developed in this treatment will remain applicable.

Two propositions underlie the present treatment. The first, the linear variation of composite properties with the concentration of components, is central to the reduction to the Linear Programming algorithm. Certain longitudinal properties have been demonstrated to follow a linear form for binary systems; it is reasonable to expect similar properties for multicomponent systems to follow linear forms. Transverse properties may deviate from linearity even in binary systems. In such cases, one has the option of utilising more involved non-linear programming techniques or taking a conservative linear estimate for composite properties. The latter choice is probably adequate for the purposes of performance optimisations.

The strain incompatible failure criterion is the second and critical proposition. Adherence to this overly conservative criterion tends to limit the possible combinations of fibres. For example, combinations of glass/boron and glass/graphite are rejected under this criterion because of the significance difference in breaking strains. Studies of the static and fatigue strengths of various combination of fibres will hopefully result in more realistic strain and strength related design criteria and lead to a more realistic constitutive equation for the strength of multicomponent systems.

In closing, it should be noted that the potential for continuously varying material properties through the control of composition introduces a strong coupling between structural and materials design. Indeed, it would appear that the effective use of multicomponent composite systems will require the integration of materials design procedures into structural design methodology.

REFERENCES

1. CHAMIS, L. C. and SENDECKYJ, G. P. *J. Composite Materials*, **2** (1968), p. 332.
2. ASHTON, J. E., HALPIN, J. C. and PETIT, P. W. *Primer on composite materials: Analysis*, Technomic Publishing Co., Stamford, Conn., 1969.
3. McCULLOUGH, R. L., *Concepts of Fiber-Resin Composites*, Marcel Dekker, Inc., New York, 1971.
4. JONES, R. M. *Mechanics of composite materials*, Scripts Book Co., Washington, DC, 1975.
5. LOOMBA, N. P. *Linear programming: An introductory analysis*, McGraw-Hill Book Co., New York, 1974.

Chapter 7

CONSTITUTIVE RELATIONSHIPS FOR HETEROGENEOUS MATERIALS

C.-T. D. Wu

Mobil Research and Development Corporation, New Jersey, USA

&

R. L. McCullough

University of Delaware, Newark, Delaware, USA

SUMMARY

Improved variational theorems are presented which provide for the subsequent development of a general constitutive relationship for effective elastic properties of heterogeneous materials: polycrystalline materials, crystalline polymers, continuous fibre-reinforced composites, discontinuous fibre-reinforced composites and particulate-reinforced systems. This formulation characterises explicitly the statistical correlation nature of heterogeneous systems and thereby reflects, inter alia, *the role of the microstructure in the determination of effective properties.*

The application of new variational theorems leads to a general bounding theory capable of constructing successively closer upper and lower bounds on the effective elastic properties by the systematic incorporation of well-defined structural descriptors. In recognition of the limited amount and restricted forms of accessible information, the theory is cast in an optimal form which isolates the roles of the dominant structural features: (i) volume fraction concentrations, (ii) orientation distribution characteristics, and (iii) statistical symmetry characteristics associated with the microstructure. This bounding theory provides the best possible bounds for each level of available information.

The bounding formula may be converted to a general constitutive relationship by merging the remaining (unspecified) structural features into the specification of a reference elasticity. In contrast to the existing combining theories which were designed for, and are consequently limited to, the consideration of a few special cases, the present theory characterises a broad spectrum of microstructures. Consequently, this theory contains various bounding treatments and micromechanical constitutive models as special cases and this provides a concise theoretical framework for the analysis of effective properties of general heterogeneous systems.

I. INTRODUCTION

The prediction of effective elastic moduli is a central problem underlying the development of quantitative structure–property relationships for a wide variety of heterogeneous materials, viz., multi-component composites, polycrystalline aggregates, and partially crystalline polymers. In principle, the exact effective moduli can be calculated only if a complete description of the microstructure and properties of the individual phases involved is provided. For systems with perfectly ordered packing (such that their microstructures can be completely described by some periodic functions) the determination of effective moduli becomes a simple computational task that can be carried out with the aid of high speed digital computers. However, most heterogeneous systems do not possess a deterministic microstructure. The stochastic nature of the microstructure of most of the important heterogeneous systems can best be described in terms of statistical characteristics. For these systems, it is reasonable to anticipate that a complete structural description will require an infinite amount of information. This situation gives rise to two problems. First, the accessibility of structural information must be evaluated and additional structural information (that can be made available through new experimental observations) identified. Secondly, combining rules (con- stitutive relationships for the prediction of effective properties) based only on a partial (finite) description of the crucial features of the microstructure must be developed and evaluated. Clearly, these two problems cannot be considered separately. The structural information must be in a form amenable to use in the combining rules while the combining rules should be expressed in terms of accessible structural descriptors.

Two approaches are commonly used to develop combining rules based on a partial description of the microstructure, viz., the modelling and the bounding methods. The modelling approach is usually based on assumptions made to approximate the stress (or strain) distribution and/or the microstructure in the system. These assumptions have not been justified beyond intuitive arguments. This approach, albeit arbitrary, does provide an unique estimation of the effective moduli. In order to avoid such assumptions, one must sacrifice the uniqueness of the solution by resorting to a bounding approach which provides only the upper and lower limits of the effective behaviour. The bounding approach possesses the advantage that the bounds are always valid no matter what the unknown (or inaccessible) part of the structural information happens to be. In the fortunate case when the calculated upper and lower bounds are reasonably close to each other, the bounding approach is obviously superior and immediately reveals that the unknown (or inaccessible) structural information has little influence on the particular property of the system under consideration. However, for many cases, the two bounds may be widely apart. Under this circumstance, the modelling approach has been regarded by many workers to be the only means to obtain reasonable estimates of the effective properties.

In the event that the bounds are too far apart and are the best possible bounds for the available information, the actual effective modulus must be some value intermediate between the bounds. Modelling estimates, based on the same amount of information, may artificially assign a specific intermediate value to the effective modulus. It is highly likely that this specific assignment bears the same order of magnitude of uncertainty as the separation between the bounds. Alternatively, it can be argued that models based on reasonable assumptions should resemble physical realities and, consequently, should yield reasonable estimates. Clearly, the justification of a model must be based on an experimental verification which presumes that the appropriate experimental variables have been identified and interrogated by the experimental technique. Most of the modelling treatments deal with the exact solution of some simplified elastic problem. Consequently, they have been applied to systems of rather simple phase geometry and structure.

In view of the limitations of the modelling approach, we propose to develop a general bounding theory capable of systematically accepting additional structural information to decrease the difference between the upper and lower bounds. It is expected that the resulting bounding theory should serve to order the relative importance of various structural features of the microstructure and thereby provides supplemental support for modelling estimates. Such a bounding approach can be formulated in terms of variational principles. Within the framework of variational calculus, the solution of a physical problem is treated as the stationary trajectories of certain variables, which correspond to the extreme, say, minimum, of some objective integral, with those variables possibly subject to a set of constraints. Any deviation from the stationary trajectories, which still satisfies the constraints, will result in an objective value above the stationary value (the minimum). In virtue of this behaviour, variational methods have been widely used in the following two types of problems. For systems which can be described completely and of which the exact solutions are not possible because of mathematical difficulties, variational methods are commonly used as a scheme to determine the best approximate solution among a set of proposed guess or trial solutions. For systems which cannot be described completely and of which the exact solutions are not possible because of the lack of information, variational methods can be used to generate upper and lower bounds on the stationary value of the objective so that limiting behaviour can be identified. The determination of effective moduli of heterogeneous systems belongs to the second type of problem, since only a partial description of the microstructure is usually available.

Common procedures of finding bounds involve the following steps:

(i)   Construct the variational form of the governing equation(s).

(ii) Transform the original minimisation (maximisation) problem into an equivalent maximisation (minimisation) problem. (By equivalent, we mean that both problems result in the same stationary solution.) The reciprocal transformation in the variational calculus is a powerful and commonly used technique for conducting this step.

(iii) Select a set of trial functions as approximations to the stationary trajectory for every variational variable involved. These functions should be made admissible to all the constraints required by the variational problem.

(iv) Evaluate the value of the objective integral corresponding to each of the trial functions and determine which one results in a value closest to the stationary value. Thus, for a minimum problem, the trial function that gives the smallest objective value should be the best among the set of trial functions selected.

(v)   Application of steps (iii) and (iv) to each of the equivalent maximum and minimum problems provides bounds on the stationary value of the objective from both above and below.

Two points should be emphasised. For problems of the first type, the trial functions may be made as elaborate as desired as long as they satisfy the constraints and as long as the mathematics involved can be handled. This is not, however, the case for problems of the second type. Under the circumstance that only a partial description of the system is available, the choice of trial functions is no longer arbitrary since the correspondent objective must be evaluated with the amount of information provided. There is an additional significant difference in nature between these two types of problems. For problems of the first type, the stationary value of the objective can in principle be approached indefinitely by bringing the trial function closer and closer to the actual solution. Indeed, in this case, the least upper bound and the greatest lower bound coincide with the stationary value itself. In contrast, for problems of the second type, there exist, for a given amount of information, the least upper bound and the greatest lower bound that remain distinct. The separation between these two bounds can be viewed as a measure of the influence exercised on the objective by the unknown part of the information. Consequently, the application of variational principles to problems of the second kind directs attention to the following tasks: how to construct the best possible bounds from a given amount of information, and how to improve the bounds by incorporating additional accessible information.

It can be anticipated that the minimum available information will consist of (i) the individual phase moduli, (ii) the orientation distributions of the anisotropic phases, and (iii) the gross volume fraction concentrations of the different phase media. Based on this information, the primitive Voigt- and Reuss-type upper and lower bounds on the effective elastic tensor may be constructed. The elements of this tensor characterise the overall elastic response of the heterogeneous system. Unfortunately, these primitive bounds are usually so far apart that it is necessary to take a more advanced approach for the bounding analysis.

There have been many articles dealing with the technique of finding improved bounds on the effective moduli. The most fruitful approach appears to be the treatments developed by Hashin and Shtrikman[1-3] and by Walpole.[4-6] Although not readily recognised, both theories are basically equivalent, even though they were derived from different (yet correspondent) variational principles. The feature that leads to the construction of improved bounds in these two treatments is an optimised choice of a trial stress function. As noted by these authors, this trial function is admissible to the set of constraints imposed on the stress–strain field only under the conditions that the material (i) behaves isotropically on the whole and (ii) is statistically homogenous as well as isotropic. Consistent with these constraints, these treatments can only be applied to heterogeneous systems of statistical isotropy (and continuous fibre-reinforced materials of statistical transverse isotropy). Many solid materials of engineering interest do not belong to these categories; one important example is the class of partially crystalline polymers. Preferred orientation of a constituent phase and complicated morphology are frequently observed. For these systems, the Voigt and Reuss averages are the only valid bounding theories currently available. However, any drastic contrast between phase properties will make these bounds uselessly far apart. In view of these limitations, a more general analysis is clearly needed.

The purpose of the current treatment is to develop a more general variational approach which can be used to systematically incorporate structural information into bounding theories. A rigorous application of variational principles will be introduced, new variational theorems will be derived, and an improved general bounding theory will be developed. This bounding theory will be further interrogated to establish reductions to a general constitutive relationship. The resulting formulation contains various existing variational treatments and most of the micromechanical models as special cases and provides the mechanism for the systematic incorporation of well-defined structural parameters. The extent to which bounds are tightened by the introduction of additional structural information will provide a clear ordering of the roles of various structural features in determining the mechanical performance.

II.   CLASSICAL VARIATIONAL METHODS

Classical variational principles have served as the basis for most bounding theories. Although these principles are treated in detail in several standard texts,[7-9] it will be convenient to summarise briefly the elements of this method in order to provide the appropriate background and definition of terms used in the current treatment. Furthermore, it will be useful to cast the classical treatment into a form which provides for a clear transition to the new variational theorems developed in the next section.

A.   Variational Formulation for Heterogeneous Systems

The principle of minimum potential energy states that for an initially stress-free linear elastic medium subject to some prescribed surface displacement which induces infinitesimal deformation of the body, the equilibrium stress and strain fields within the medium should be given by those which result in the minimum of the strain energy functional defined as

Objective:
$$U_0 = \tfrac{1}{2} \int \sigma_{ij} \varepsilon_{ij} \, dV \qquad (1)$$

subject to subsidiary conditions:

Constitutive:
$$\sigma_{ij} = C_{ijkl} \varepsilon_{kl} \qquad (2)$$

Compatibility:
$$\varepsilon_{ij} = \tfrac{1}{2}(u_{i,j} + u_{j,i}) \qquad (3)$$

Boundary Condition:
$$u_i(S) = u_i^s(S) \qquad (4)$$

where σ_{ij} and ε_{ij} are the Cartesian stress and infinitesimal strain tensor fields, respectively. C_{ijkl} is the linear elastic field of the system under consideration. The u_is denote the displacement vector field; $u_i^s(S)$ is the prescribed displacement on the surface of the body denoted symbolically by S. The subscript j denotes the spacial derivative in the j direction. The integration in eqn. (1) is taken over all the volume elements bounded by surface S.

It can be shown, through the method of Lagrangian multiplier, that the stationary solution of the above variational problem corresponds to the equilibrium condition:

Equilibrium:
$$\sigma_{ij,j} = 0 \qquad (5)$$

Equation (5) together with eqns. (2) to (4) gives the differential equation version of the original variational problem.

Some precautions have to be made in order to apply the above variational principle and other alternative forms which will be developed later to heterogeneous media, because of the discontinuities in their elasticity and strain fields across the phase boundaries. These discontinuities will not give rise to any problem if it is required that the displacement field is always continuous. It is expected that the cohesion between contiguous phases is strong enough to resist small deformation without slip at the interface. Consequently, we restrict our consideration only to heterogeneous systems defined as follows:

(i) A heterogeneous system is a mixture of several media (which will be called phases) of distinct physical properties.

(ii)   Each medium forms regions (which will be called phase regions) that are large enough to be regarded as continua.

(iii) All phase regions are firmly bonded together such that upon an external load the displacement field is always continuous.

(iv)   Without the external load, the system is free of any internal stress.

Under the above assumptions, it is shown in Appendix 1 that by modifying the subsidiary condition equation (3) as $\varepsilon_{ij} = \frac{1}{2}(u_{i,j} + u_{j,i})$ within phase regions and the u_is are continuous at phase interfaces, the principle of minimum potential energy can be directly applied to heterogeneous systems and predicts the correct equilibrium condition, viz., eqn. (5) within every phase region with interfacial forces balanced. For simplicity of notation in later treatments, once equation (3) is written for a heterogeneous system, it automatically implies the above modification.

The variational approach has the advantage over its differential equation version in the sense that for cases when exact solutions are not attainable, the variational method will help to make the best choice among a set of testing or trial approximating solutions. It also possesses the feature that under a certain transformation of an original minimisation problem, a maximisation problem, which carries the same stationary solution, is obtained. This important property associated with the transformations provides a powerful tool for establishing both the upper and the lower limits on the stationary value of the objective integral and, consequently, on the overall property of the system that is characterised by the objective.

B.   Transformations of Variational Problems

The transformations of a variational problem are accomplished by incorporating some or all of the subsidiary and/or stationary conditions into the objective integral through the method of Lagrangian multiplier. It can be anticipated that several alternative transformations may be applied to a given variational problem. In order to preserve the stationary character of the original problem, certain transformation rules must always be observed. A detailed discussion of various transformations is given in the classical textbook by Courant and Hilbert.[10] Here it is sufficient to review some important concepts related to the present work.

In addition to the capability of converting a minimum problem to an equivalent maximum problem, transformation may sometimes be used to simplify a variational problem. Among many possible transformations is the reciprocal form (also known as the involuntary transformation) which is of special interest both theoretically and practically because of its involutorily symmetric character. The reciprocal form of a variational problem is obtained by interchanging the role of the subsidiary conditions with that of the stationary conditions through the use of Lagrangian multipliers. This procedure will be illustrated in the following paragraphs by constructing the reciprocal form for the principle of minimum potential energy.

According to the principle of minimum potential energy, the equilibrium strain energy of a system under prescribed surface displacement corresponds to the minimum value of the objective integral. Consequently, any approximating stress (or strain) field deviated from the stationary solution will lead to an objective value above the equilibrium strain energy. Thus an upper bound on the equilibrium strain energy of the system is obtained. For the purpose of bracketing the overall response, a lower bound is also required. This requirement dictates the need for another variational formulation which predicts the same equilibrium strain energy as a maximum. This correspondent maximum problem can often be found by appropriate transformations.

The reciprocal transformation will be considered first. For the variational problem defined by eqns. (1) to (4), a slight change in form of the objective will cause the associated reciprocal problem to assume a simpler form. With the help of eqn. (2), eqn. (1) may be written as

$$U_0 = \tfrac{1}{2} \int \sigma_{ij} S_{ijkl} \sigma_{kl} \, dV \qquad (6)$$

where S_{ijkl} represents the compliance field which satisfies the following relation at every point in the system:

$$S_{ijmn} C_{mnkl} = I_{ijkl} \qquad (7)$$

with I_{ijkl} defined as

$$I_{ijkl} = \tfrac{1}{2}(\delta_{ik}\delta_{jl} + \delta_{il}\delta_{jk}) \qquad (8)$$

where δ_{ij} is the Kronecker delta. For a symmetric tensor $t_{ij} = t_{ji}$, I_{ijkl} serves as the unit operator in the sense that $I_{ijkl}t_{kl} = t_{ij}$.

To find the reciprocal form of the variational problem of minimising the objective, eqn. (6), subject to subsidiary conditions (eqns. (2) to (4)), the subsidiary conditions are first incorporated into the objective through Lagrangian multipliers β_{ij}, γ_{ij}, and λ_{ij}. This gives

$$2U = \int [\sigma_{ij} S_{ijkl} \sigma_{kl} + \beta_{ij}(\sigma_{ij} - C_{ijkl}\varepsilon_{kl}) + \gamma_{ij}(2\varepsilon_{ij} - u_{i,j} - u_{j,i})] \, dV$$
$$+ \int \lambda_{ij}(u_i - u_i^s)n_j \, dS \qquad (9)$$

where n_j is the jth component of the normal vector at surface element dS.

With the help of the Divergence Theorem, eqn. (9) can be written as:

$$2U = \int [\sigma_{ij}S_{ijkl}\sigma_{kl} + \beta_{ij}(\sigma_{ij} - C_{ijkl}\varepsilon_{kl}) + 2\gamma_{ij}\varepsilon_{ij} + (\gamma_{ij} + \gamma_{ji})_{,j}u_i]\,dV$$
$$+ \int [\lambda_{ij}u_i - \lambda_{ij}u_i^s - (\gamma_{ij} + \gamma_{ji})u_i]n_j\,dS \qquad (10)$$

The objective given by eqn. (10) is free from constraints on the variables σ_{ij}, ε_{ij}, u_i, β_{ij}, γ_{ij}, and λ_{ij}. By equating its variation with respect to these variables to zero, the stationary conditions may be generated, among which are the original subsidiary conditions as well as the following:

$$-C_{ijkl}\beta_{kl} + (\gamma_{ij} + \gamma_{ji}) = 0 \qquad (11)$$

$$2S_{ijkl}\sigma_{kl} + \beta_{ij} = 0 \qquad (12)$$

$$(\gamma_{ij} + \gamma_{ji})_{,j} = 0 \qquad (13)$$

$$\lambda_{ij}(S) - (\gamma_{ij}(S) + \gamma_{ji}(S)) = 0 \qquad (14)$$

Elimination of the multipliers from these equations will lead to the stationary condition equation (5).

Equations (2) to (4) and (11) to (14) are the stationary conditions of the constraint-free objective equation (10), which can be transformed by taking some of its stationary relations as subsidiary conditions and using these relations to eliminate unnecessary variables. If eqns. (2) to (4) are taken as subsidiary conditions, eqn. (10) can be simplified to eqn. (1) or eqn. (6) to recover the original problem. If, instead, eqns. (11) to (14) are taken as subsidiary conditions, the reciprocal problem is obtained, which after the elimination of the multipliers reads: The reciprocal objective defined as

$$U_r = \int \sigma_{ij}u_i^s n_j\,dS - \tfrac{1}{2}\int \sigma_{ij}S_{ijkl}\sigma_{kl}\,dV \qquad (15)$$

subject to the subsidiary condition (5) is stationary when equations (2) to (4) are satisfied. This treatment illustrates that the reciprocal problem is indeed equivalent to the original problem in the sense that the solutions of both problems are governed by the same set of differential equations, viz. (2) to (5). The difference is that the roles of these equations as stationary or subsidiary conditions are now reversed. In contrast to the original minimum problem, the reciprocal objective predicts the equilibrium strain energy as a maximum due to the negativeness of its second variation:

$$\partial^2 U_r = -\int \partial\sigma_{ij}S_{ijkl}\partial\sigma_{kl}\,dV < 0$$

since S_{ijkl} should always be positive definite.

This reciprocal variational problem is known as the principle of complementary energy for systems subject to prescribed surface displacement. Since the reciprocal transformation provides an equivalent maximum problem, the equilibrium strain energy of the system can now be bounded from both above and below.

C. Bounding

As noted in the preceding review, the classical variational treatments provide a minimum principle and an equivalent maximum principle. Bounds will be developed based on these principles in order to (i) illustrate the bounding procedures outlined in the previous section, (ii) explore the limitations of the classical variational principles, and (iii) identify the additional structural information required for further improving the bounds.

Usually, sufficient information is accessible to perform the volume average of the elastic field. Without this information, the effective properties are trivially bounded by the 'stiffest' possible phase property from above and by the 'softest' from below. Thus, the information required for volume averaging is regarded as the essential minimum amount of information. For a multiphase system, this includes (i) all the individual phase moduli, (ii) gross volume fractions of various phases, and (iii) the orientation distributions of anisotropic phase components.

With the minimum information provided, consider a heterogeneous system of unit volume subject to the prescribed surface displacement:

$$u_i(S) = \varepsilon_{ij}^0 x_j(S) \tag{16}$$

where $\varepsilon_{ij}^0 = \varepsilon_{ji}^0$. Note that the imposed surface displacement was specified in such a way that if the system happens to be homogeneous, it will undergo homogeneous deformation with constant strain ε_{ij}^0. The effective elasticity C_{ijkl}^* of the system is defined as the elasticity of a homogeneous material which stores the same amount of strain energy as the actual system under the same surface displacement. In other words, C_{ijkl}^* is defined through the following relation:

$$\tfrac{1}{2}\varepsilon_{ij}^0 C_{ijkl}^* \varepsilon_{kl}^0 = \langle U^s \rangle \tag{17}$$

where $\langle U^s \rangle$ is the equilibrium strain energy per unit volume of the actual heterogeneous system.

Bounds on the equilibrium strain energy, and consequently on the effective elasticity, of a heterogeneous system are generated by following the bounding procedures laid out in Section I. The classical variational principles provide for the first two steps:

(i) The minimum problem: The objective

$$U_0 = \tfrac{1}{2} \int \sigma_{ij} \varepsilon_{ij} \, \mathrm{d}V$$

subject to constraints:

$$\sigma_{ij} = C_{ijkl} \varepsilon_{kl}$$

$$\varepsilon_{ij} = \tfrac{1}{2}(u_{i,j} + u_{j,i})$$

$$u_i(S) = \varepsilon_{ij}^0 x_j(S)$$

should be minimised to give the equilibrium strain energy.

(ii) The maximum problem: The objective

$$U_r = \int \sigma_{ij} u_i n_j \, \mathrm{d}S - \tfrac{1}{2} \int \sigma_{ij} S_{ijkl} \sigma_{kl} \, \mathrm{d}V$$

subject to the constraint:

$$\sigma_{ij,j} = 0$$

should be maximised to give the equilibrium strain energy.

The choice of the trial fields must be consistent with the subsidiary conditions and the amount of information available. For the current case, a reasonable guess is that ε_{ij}^t is constant for the minimum problem and σ_{ij}^t is constant for the maximum problem, where the superscript t denotes 'trial'. In order to satisfy all the subsidiary conditions in the minimum problem,

$$\varepsilon_{ij}^t = \varepsilon_{ij}^0 \qquad \text{and} \qquad \sigma_{ij}^t = C_{ijkl} \varepsilon_{kl}^0$$

is the only choice. For the maximum problem, $\sigma_{ij}^t = \text{constant}$ certainly satisfies the constraint. Accordingly, the following trial functions are obtained:

(iii) For the minimum problem,

$$\varepsilon_{ij}^t = \varepsilon_{ij}^0 \qquad \text{and} \qquad \sigma_{ij}^t = C_{ijkl} \varepsilon_{kl}^0$$

For the maximum problem,

$$\sigma_{ij}^t = \text{constant}$$

The evaluation of the correspondent objective as well as the determination of the best trial function is conducted in step (iv). For the minimum problem, the trial function has been fixed by the boundary condition. For the maximum problem, the objective corresponding to a constant stress field is given by:

$$U_r^t = \int \sigma_{ij}^t u_i n_j \, dS - \tfrac{1}{2} \int \sigma_{ij}^t S_{ijkl} \sigma_{kl}^t \, dV$$

$$= \sigma_{ij}^t \varepsilon_{ij}^0 - \tfrac{1}{2} \sigma_{ij}^t \langle S_{ijkl} \rangle \sigma_{kl}^t$$

where $\langle S_{ijkl} \rangle$ is the volume average over the compliance field. The σ_{ij}^t which gives the maximum U_r^t can easily be found by equating ∂U_r^t equal to zero. This gives

$$\varepsilon_{ij}^0 = \langle S_{ijkl} \rangle \sigma_{kl}^t$$

In summary, the best trial functions are respectively:

(iv) For the minimum problem,

$$\varepsilon_{ij}^t = \varepsilon_{ij}^0 \qquad \text{and} \qquad \sigma_{ij}^t = C_{ijkl} \varepsilon_{kl}^0 \tag{18}$$

with the correspondent U_0 given by

$$U_0^t = \tfrac{1}{2} \varepsilon_{ij}^0 \langle C_{ijkl} \rangle \varepsilon_{kl}^0 \tag{19}$$

For the maximum problem,

$$\sigma_{ij}^t = \langle S_{ijkl} \rangle^{-1} \varepsilon_{kl}^0 \tag{20}$$

with the correspondent U_r given by

$$U_r^t = \tfrac{1}{2} \varepsilon_{ij}^0 \langle S_{ijkl} \rangle^{-1} \varepsilon_{kl}^0 \tag{21}$$

Because the equilibrium strain energy, U^s, of the heterogeneous system (of unit volume) corresponds to the minimum in the minimum problem and to the maximum in the maximum problem, it can be asserted that

(v) $U_r^t \leq U^s \leq U_0^t$. From eqns. (17), (19), and (21), the effective elasticity can be bounded from both above and below as:

$$\langle S_{ijkl} \rangle^{-1} \leq C_{ijkl}^* \leq \langle C_{ijkl} \rangle \tag{22}$$

where $\mathbf{A} < \mathbf{B}$ (with $\mathbf{A}$ and $\mathbf{B}$ symmetric tensors of the same rank) means that $\mathbf{B} - \mathbf{A}$ is positive definite.

Note that eqn. (22) does not imply that every element of C_{ijkl}^* is bounded by the correspondent elements in $\langle S_{ijkl} \rangle^{-1}$ and $\langle C_{ijkl} \rangle$. Instead, it implies that the strain energy corresponding to C_{ijkl}^* is always bounded by those correspondent to $\langle S_{ijkl} \rangle^{-1}$ and $\langle C_{ijkl} \rangle$ for all ε_{ij}^0.

It is evident that for a multi-component system

$$\langle A_{ijkl} \rangle = \sum_m v_m \langle A_{ijkl}^m \rangle \tag{23}$$

where v_m is the volume fraction of the mth component and $\langle A_{ijkl}^m \rangle$ is the orientation average of the appropriate tensorial quantity associated with the mth component (e.g. $A_{ijkl} = S_{ijkl}$, C_{ijkl}, or some combination of S_{ijkl} and C_{ijkl}). Explicit forms for $\langle A_{ijkl}^m \rangle$ in terms of characteristics of orientation distributions are given in Appendix 6.

It should be noted that the upper bound corresponds to the volume average over the elasticity field, which is the averaging rule proposed by Voigt.[11] Because the elasticity field in a heterogeneous system is not uniform, a constant strain field will result in an non-equilibrium stress field. If we envisage a heterogeneous system under constant strain, the correspondent non-equilibrium stress field will drive the system in motion. It is therefore obvious that the situation of constant strain corresponds to an energy state higher than the situation of static equilibrium. This is, of course, the reason why the Voigt average gives an upper bound.

On the other hand, the lower bound is given by the inverse of the volume average over the compliance field and is exactly the averaging rule proposed by Reuss.[12] The Reuss average is derived by assuming a constant stress field, which certainly fulfils the force equilibrium condition. However, because of the heterogeneous elasticity field, the strain field is no longer compatible in the sense that the displacement field is not continuous. If we envisage a heterogeneous system under a constant stress field, there will be material overlaps and voids in the vicinities of phase boundaries. This is certainly not permitted for perfectly bonded systems. In order to adjust the displacement field so as to eliminate overlaps and voids, additional energy will be required to bend or distort the material elements to the suitable positions. For this reason, the Reuss average results in a strain energy lower than the real situation and hence serves as a lower bound.

Within the provisions of the information required for performing the volume average, the classical variational treatments provide for the construction of primitive lower and upper bounds, viz., the Reuss and Voigt averages respectively. In many cases, however, these two bounds are so far apart that bound improvement is essential in order to obtain a reasonable estimate of the effective performance.

III.   REFINED FORMULATION

A.   Variational Principles Based on an Unspecified Reference System

The classical variational principles discussed previously will be refined in this section by considering, instead of the actual system, its deviation from some unspecified reference system. The motivation for this is to introduce flexibility into the variational principle. The properties of the yet unspecified reference system will be introduced into the bounding formulation in such a way that by adjusting its properties, tighter bounds can be obtained. Methods for specifying the appropriate reference system will be developed in a subsequent section.

To avoid unnecessary complications in later analyses, the reference system will be taken as a homogeneous medium of elasticity C^0_{ijkl} subject to the surface displacement $u^s_i(S)$ prescribed for the actual system. At this point, C^0_{ijkl} is taken to be a general anisotropic elasticity tensor. The resulting displacement, strain, and stress fields of the reference system are denoted by u^0_i, ε^0_{ij}, and σ^0_{ij}, respectively. As shown in the previous section, these fields should satisfy the following relations simultaneously:

$$\sigma^0_{ij} = C^0_{ijkl}\varepsilon^0_{kl} \tag{24}$$

$$\varepsilon^0_{ij} = \tfrac{1}{2}(u^0_{i,j} + u^0_{j,i}) \tag{25}$$

$$u^0_i(S) = u^s_i(S) \tag{26}$$

$$\sigma^0_{ij,j} = 0 \tag{27}$$

The elasticity of the actual system subject to the same surface displacement $u^s_i(S)$ is assumed to be heterogeneous in general and to be representable by a space function $C_{ijkl}(\mathbf{r})$. The strain and stress fields and the associated strain energy induced by the surface displacement can in principle be determined by solving the variational problem given by eqns. (1) to (4) or the set of differential eqns. (2) to (5). However, for this treatment, attention is directed to the deviation of the actual system from the reference system. The following deviatoric quantities are thus introduced:

$$\sigma_{ij} = C^0_{ijkl}\varepsilon_{kl} + P_{ij} \tag{28}$$

$$R_{ijkl} = C_{ijkl} - C^0_{ijkl} \tag{29}$$

$$H_{ijmn}R_{mnkl} = I_{ijkl} \tag{30}$$

$$\varepsilon_{ij} = \varepsilon^0_{ij} + \varepsilon'_{ij} \tag{31}$$

$$u_i = u^0_i + u'_i \tag{32}$$

where p_{ij} is the stress polarisation tensor, which was used by Eshelby[13, 14] in the analysis of inclusion problems and was subsequently treated by Hashin et al.[1-3] as the major variable in their variational treatments.

In terms of these deviatoric quantities, the variational problem specified by eqns. (1) to (4) can be restated as:

The objective given by

$$U_0 = U^0 + \tfrac{1}{2} \int (\varepsilon'_{ij} C^0_{ijkl} \varepsilon'_{kl} + p_{ij} H_{ijkl} p_{kl}) \, dV \tag{33}$$

should be minimised, subject to the subsidiary conditions:

$$p_{ij} = R_{ijkl}(\varepsilon^0_{kl} + \varepsilon'_{kl}) \tag{34}$$

$$\varepsilon'_{ij} = \tfrac{1}{2}(u'_{i,j} + u'_{j,i}) \tag{35}$$

$$u'_i(S) = 0 \tag{36}$$

to give the equilibrium strain energy and the correspondent stress and strain fields of the actual system. The U^0 in eqn. (33) is defined as

$$U^0 = \tfrac{1}{2} \int \varepsilon^0_{ij} C^0_{ijkl} \varepsilon^0_{kl} \, dV \tag{37}$$

which is the equilibrium strain energy of the reference system as ε^0_{ij} is assumed to be the strain field that satisfies eqns. (24) to (27).

It should be noted that in the derivation of eqn. (33), eqns. (27), (35) and (36) have been used with the divergence theorem to eliminate the following term:

$$\begin{aligned}
\int \varepsilon^0_{ij} C^0_{ijkl} \varepsilon'_{kl} \, dV &= \int \sigma^0_{ij} u'_{i,j} \, dV \\
&= \int \sigma^0_{ij} u'_i n_j \, dS - \int \sigma^0_{ij,j} u'_i \, dV \\
&= 0
\end{aligned} \tag{38}$$

because $u'_i = 0$ on S and $\sigma^0_{ij,j} = 0$ in V. It should also be pointed out that eqn. (34) has been used to transform the objective into the form of eqn. (33). As shown in the previous section, such transformations do not change the stationary character of the variational problem and will yield a better form for the reciprocal problem.

The reciprocal transformation of the variational problem specified by eqns. (33) to (36) will now be considered. Incorporation of the subsidiary conditions through Lagrangian multipliers α_{ij}, β_{ij}, and λ_{ij} converts eqn. (33) to:

$$\begin{aligned}
U = U^0 + \tfrac{1}{2} \int [\varepsilon'_{ij} C^0_{ijkl} \varepsilon'_{kl} &+ p_{ij} H_{ijkl} p_{kl} + \alpha_{ij}(p_{ij} - R_{ijkl}\varepsilon^0_{kl} - R_{ijkl}\varepsilon'_{kl}) \\
&+ \beta_{ij}(2\varepsilon'_{ij} - u'_{i,j} - u'_{j,i})] \, dV + \int u'_i \lambda_{ij} n_j \, dS
\end{aligned} \tag{39}$$

or equivalently

$$U = U^0 + \tfrac{1}{2} \int [\varepsilon'_{ij} C^0_{ijkl} \varepsilon'_{kl} + p_{ij} H_{ijkl} p_{kl} + \alpha_{ij} p_{ij} - \alpha_{ij} R_{ijkl} \varepsilon^0_{kl}$$
$$- \alpha_{ij} R_{ijkl} \varepsilon'_{kl} + 2\beta_{ij} \varepsilon'_{ij} + u'_i (\beta_{ij} + \beta_{ji})_{,j}] \, \mathrm{d}V$$
$$+ \int [\lambda_{ij} u'_i - (\beta_{ij} + \beta_{ji}) u'_i] n_j \, \mathrm{d}S \qquad (40)$$

Taking the variation of U and setting it equal to zero generates the relations (34), (35) and (36), and the following stationary conditions:

$$C^0_{ijkl} \varepsilon'_{kl} - R_{ijkl} \alpha_{kl} + 2\beta_{ij} = 0 \qquad (41)$$

$$H_{ijkl} p_{kl} + \alpha_{ij} = 0 \qquad (42)$$

$$(\beta_{ij} + \beta_{ji})_{,j} = 0 \qquad (43)$$

$$\lambda_{ij}(S) - \beta_{ij}(S) - \beta_{ji}(S) = 0 \qquad (44)$$

The reciprocal transformation can then be constructed by incorporating eqns. (41) to (44) with (40). After eliminating the multipliers, the reciprocal problem takes on the following form:

The reciprocal objective:

$$U_r = U^0 - \int (\tfrac{1}{2} p_{ij} H_{ijkl} p_{kl} + \tfrac{1}{2} \varepsilon'_{ij} C^0_{ijkl} \varepsilon'_{kl} - p_{ij} \varepsilon^0_{ij}) \, \mathrm{d}V \qquad (45)$$

subject to the subsidiary condition:

$$(C^0_{ijkl} \varepsilon'_{kl} + p_{ij})_{,j} = 0 \qquad (46)$$

is stationary at

$$p_{ij} = R_{ijkl}(\varepsilon^0_{kl} + \varepsilon'_{kl}) \qquad (47)$$

$$\varepsilon'_{ij} = \tfrac{1}{2}(u'_{i,j} + u'_{j,i}) \qquad (48)$$

$$u'_i(S) = 0 \qquad (49)$$

and the equilibrium strain energy of the actual system, denoted by U^s, is given by the stationary value of the reciprocal objective.

Further transformations are required for the following reasons:

(i) A unique relationship between p_{ij} and ε'_{ij} is desired because substitution of this relationship into the objective (45) provides a constraint-free objective of only one variable. Equation (46) alone does not determine an unique solution of ε'_{ij} in terms of p_{ij}. Some boundary condition and the compatibility of ε'_{ij} field are also required.

(ii)   An examination of the second variation of the reciprocal objective shows that a choice for C^0_{ijkl}, such that R_{ijkl} is positive definite, converts the reciprocal problem to a maximum problem. Alternatively, it can be made a minimum problem by a choice of C^0_{ijkl} such that R_{ijkl} is negative definite and by restricting p_{ij} and ε'_{ij} to satisfy not only eqn. (46) but also eqns. (48) and (49).

Accordingly, the reciprocal problem is further transformed by demanding two more subsidiary conditions, viz., eqns. (48) and (49). In order to obtain the transformed objective, eqn. (45) is first written as:

$$U_r = U^0 - \int [\tfrac{1}{2}p_{ij}H_{ijkl}p_{kl} - p_{ij}\varepsilon^0_{ij} + \tfrac{1}{2}\varepsilon'_{ij}(t_{ij} - p_{ij})]\,dV \qquad (50)$$

where

$$t_{ij} = C^0_{ijkl}\varepsilon'_{kl} + P_{ij} \qquad (51)$$

and from (46),

$$t_{ij,j} = 0 \qquad (52)$$

With the newly added constraints, the following term may be eliminated:

$$\int \varepsilon'_{ij}t_{ij}\,dV = \int t_{ij}u'_i n_j\,dS - \int t_{ij,j}u'_i\,dV = 0 \qquad (53)$$

because of eqns. (49), (52), and the divergence theorem. Consequently, the transformed reciprocal objective can be reduced to the following form:

$$U_h = U^0 - \int (\tfrac{1}{2}p_{ij}H_{ijkl}p_{kl} - p_{ij}\varepsilon^0_{ij} - \tfrac{1}{2}p_{ij}\varepsilon'_{ij})\,dV \qquad (54)$$

The above objective potential was first derived by Hashin[15] through a somewhat different and less tractable approach. It should be pointed out that this potential can also be derived directly by incorporating eqns. (41) to (44) together with (35) and (36) into the objective (40). In this sense, U_0 and U_h can be regarded to be reciprocal to each other with respect to the conditions of eqns. (46) and (47) only.

The results lead to the following theorem:

Theorem 1: The objective potential U_h defined in eqn. (54), subject to the subsidiary conditions (eqns. (46), (48), and (49)) is stationary when eqn. (47) is satisfied. The equilibrium strain energy of the actual system is given by U^s_h, the stationary value of U_h, which satisfies the following inequalities:

$$U^s_h \geq U_h \qquad \text{when } R_{ijkl} \text{ is positive definite} \qquad (55)$$

while

$$U^s_h \leq U_h \qquad \text{when } R_{ijkl} \text{ is negative definite} \qquad (56)$$

for U_h resulting from admissible p_{ij} and ε'_{ij} fields that satisfy the subsidiary conditions. (The proof is given in Appendix 2.)

Theorem 1 was originally presented by Hashin and Shtrikman.[1] It possesses the highly desirable feature that by adjusting the reference system it can generate both upper and lower bounds of the equilibrium strain energy. Unfortunately, it requires that all the trial approximation fields for p_{ij} and ε'_{ij} have to be made admissible to the subsidiary conditions before they can be substituted into the objective potential for the estimation of bounds. This difficulty can be eliminated by constructing a constraint free variational problem with an attendant refinement of Theorem 1.

B.   Constraint Free Formulation

A close examination reveals that if p_{ij} is treated as an independent variable tensor function, the ε'_{ij} field can be completely determined by the subsidiary conditions (eqns. (46), (48), and (49)). These three equations can be regarded as a static elastic problem for a system of homogeneous elasticity, C^0_{ijkl}, subject to homogeneous boundary condition and body force field, $p_{ij,j}$. The resulting strain field ε'_{ij} can in general be expressed in terms of $p_{ij,j}$ through an integral operator with the Green's tensor function[16] as a weighting factor. The ε'_{ij} so determined for a given p_{ij} field satisfies the subsidiary conditions automatically. Substitution of this relation into the objective (eqn. (54)) to eliminate the presence of ε'_{ij} yields a constraint free objective with only one variable, viz., p_{ij}.

To acquire this functional relationship between ε'_{ij} and p_{ij}, eqn. (46) is revised with the help of eqn. (48) to the following form:

$$C^0_{ijkl}u'_{k,lj} + p_{ij,j} = 0 \tag{57}$$

The u'_i can be written in terms of $p_{ij,j}$ once the inverse of the differential operator,

$$C^0_{ijkl}\frac{\partial}{\partial x_l}\frac{\partial}{\partial x_j}$$

with homogeneous boundary condition (eqn. (49)) has been obtained. This inverse operator can be constructed by properly operating with the Green's tensor function $g^0_{mi}(\mathbf{r}, \mathbf{r}')$, introduced through the following integral relation:

$$\int g^0_{mi}(\mathbf{r}, \mathbf{r}')C^0_{ijkl}u'_{k,lj}(\mathbf{r}')\,\mathrm{d}V' = \int g^0_{mi}(\mathbf{r}, \mathbf{r}')C^0_{ijkl}u'_{k,l}(\mathbf{r}')n_j\,\mathrm{d}S'$$

$$- \int C^0_{ijkl}u'_k(\mathbf{r}')\frac{\partial}{\partial x'_j}g^0_{mi}(\mathbf{r}, \mathbf{r}')n_l\,\mathrm{d}S' + \int u'_k(\mathbf{r}')C^0_{ijkl}\frac{\partial^2}{\partial x'_l\,\partial x'_j}g^0_{mi}(\mathbf{r}, \mathbf{r}')\,\mathrm{d}V' \tag{58}$$

where the integrations are taken over the volume and the surface of the system with respect to the space vector $\mathbf{r}'$. The divergence theorem was used to obtain this relation.

The definition of the Green's tensor function may be introduced by demanding it satisfy the following equation:

$$C^0_{ijkl} \frac{\partial^2}{\partial x'_l \partial x'_j} g^0_{mi}(\mathbf{r}, \mathbf{r}') + \delta_{mk}\delta(\mathbf{r} - \mathbf{r}') = 0 \tag{59}$$

and the boundary condition:

$$g^0_{mi}(\mathbf{r}, \mathbf{r}'(S)) = 0 \tag{60}$$

where $\mathbf{r}'(S)$ means that $\mathbf{r}'$ is on the surface S and $\delta(\mathbf{r})$ is the Dirac delta function.

From the definition of $g^0_{mi}(\mathbf{r},\mathbf{r}')$ and the boundary condition of u'_i given by eqn. (49), eqn. (58) becomes

$$\int g^0_{mi}(\mathbf{r}, \mathbf{r}')C^0_{ijkl}u'_{k,lj}(\mathbf{r}')\,\mathrm{d}V' = -u'_m(\mathbf{r}) \tag{61}$$

From the above equation, it is noted that the integral operator with $g^0_{mi}(\mathbf{r},\mathbf{r}')$ as its kernel serves as the inverse of the differential operator, $C^0_{ijkl}(\partial^2/\partial x'_l\,\partial x'_j)$, with the homogeneous boundary condition. By operating the inverse operator on eqn. (57), the u'_i may be related to $p_{ij,j}$ through

$$u'_m(\mathbf{r}) = \int g^0_{mi}(\mathbf{r}, \mathbf{r}')p_{ij,j}(\mathbf{r}')\,\mathrm{d}V' \tag{62}$$

It can readily be shown that this relationship between u'_m and $p_{ij,j}$ indeed satisfies the set of subsidiary conditions imposed on the stress and strain fields. Before proceeding to substitute eqn. (62) into the objective (eqn. (54)), it is pertinent to discuss certain properties of the Green's tensor function. The formal differential operator $C^0_{ijkl}(\partial^2/\partial x_l\,\partial x_j)$ with homogeneous boundary condition is self-adjoint.[17] That is to say, for some functions $v(\mathbf{r})$ and $w(\mathbf{r})$ satisfying the condition that $v = w = 0$ when $\mathbf{r}$ is on S, then

$$\int v(\mathbf{r})C^0_{ijkl}w_{,lj}(\mathbf{r})\,\mathrm{d}V = \int w(\mathbf{r})C^0_{ijkl}v_{,lj}(\mathbf{r})\,\mathrm{d}V \tag{63}$$

Substituting $g^0_{nk}(\mathbf{r}', \mathbf{r})$ for $v(\mathbf{r})$ and $g^0_{mi}(\mathbf{r}'', \mathbf{r})$ for $w(\mathbf{r})$ in eqn. (63) and summing over indices k and i yields

$$\int g^0_{nk}(\mathbf{r}', \mathbf{r})C^0_{ijkl} \frac{\partial^2}{\partial x_l\,\partial x_j} g^0_{mi}(\mathbf{r}'', \mathbf{r})\,\mathrm{d}V = \int g^0_{mi}(\mathbf{r}'', \mathbf{r})C^0_{ijkl} \frac{\partial^2}{\partial x_l\,\partial x_j} g^0_{nk}(\mathbf{r}', \mathbf{r})\,\mathrm{d}V \tag{64}$$

Using the definition of g_{ij}^0, i.e., eqn. (59), and the symmetry of C_{ijkl}^0, eqn. (64) reduces to

$$\int g_{nk}^0(\mathbf{r}',\mathbf{r})\delta_{mk}\delta(\mathbf{r}'' - \mathbf{r})\,\mathrm{d}V = \int g_{mi}^0(\mathbf{r}'',\mathbf{r})\delta_{ni}\delta(\mathbf{r}' - \mathbf{r})\,\mathrm{d}V$$

This implies the following symmetry property for the Green's tensor function:

$$g_{nm}^0(\mathbf{r}',\mathbf{r}'') = g_{mn}^0(\mathbf{r}'',\mathbf{r}') \tag{65}$$

Substitution of eqn. (62) into the objective (eqn. (54)) eliminates the presence of ε_{ij}'. Note that the last integral term in (54) can also be written as:

$$\tfrac{1}{2}\int \varepsilon_{ij}' p_{ij}\,\mathrm{d}V = \tfrac{1}{2}\int u_{i,j}' p_{ij}\,\mathrm{d}V = -\tfrac{1}{2}\int u_i' p_{ij,j}\,\mathrm{d}V \tag{66}$$

because of $u_i' = 0$ on S. Combining eqns. (54), (62), and (66) yields the desired constraint-free problem:

Theorem 2: The objective

$$U_p = U^0 - \tfrac{1}{2}\int p_{ij}H_{ijkl}p_{kl}\,\mathrm{d}V + \int p_{ij}\varepsilon_{ij}^0\,\mathrm{d}V$$
$$- \tfrac{1}{2}\iint p_{ij,j}(\mathbf{r})g_{ik}^0(\mathbf{r},\mathbf{r}')p_{kl,l}(\mathbf{r}')\mathrm{d}V\,\mathrm{d}V' \tag{67}$$

is stationary with respect to the variable $p_{ij}(\mathbf{r})$ at:

$$H_{ijkl}p_{kl}(\mathbf{r}) = \varepsilon_{ij}^0 + \int \frac{1}{2}\left[\frac{\partial}{\partial x_j}g_{ik}^0(\mathbf{r},\mathbf{r}') + \frac{\partial}{\partial x_i}g_{jk}^0(\mathbf{r},\mathbf{r}')\right]p_{kl,l}(\mathbf{r}')\,\mathrm{d}V' \tag{68}$$

and the equilibrium strain energy U_p^s (the stationary value of U_p) is given by:

$$U_p^s = U^0 + \tfrac{1}{2}\int \varepsilon_{ij}^0 p_{ij}^s\,\mathrm{d}V \tag{69}$$

where p_{ij}^s denotes the stationary solution of p_{ij} field, which satisfies eqn. (68). U_p^s also satisfies the following inequalities:

$$U_p^s \geq U_p \qquad \text{when } R_{ijkl} \text{ is positive definite} \tag{70}$$

while

$$U_p^s \leq U_p \qquad \text{when } R_{ijkl} \text{ is negative definite} \tag{71}$$

Note that the stationary condition (eqn. (68)) is exactly the same as eqn. (47) because the integral term in eqn. (68) is nothing more than the admissible ε_{ij}' field in eqn. (47). Equation (68) alone can also be viewed as the integral equation version of the set of differential equations (46) to (49). This relationship is an integral equation of the second kind[10] and can in principle be solved by the scheme of successive approximation provided the sequence of the successive solutions converges. It can also be solved by resorting to its differential equation version.

Equation (69) can be obtained by direct substitution of eqn. (68) into (67). This relation was first derived by Eshelby.[13] Its simple form enhances the practicability of analysing stress-strain problems through stress polarisation tensor fields.

The superiority of Theorem 2 over Theorem 1 is obvious. First, there are no constraints imposed on the variation variable p_{ij}. Any trial field for p_{ij} will be admissible to the variational problem. Second, the last double integral term in the objective (eqn. (67)) characterises explicitly the correlation nature of the stress field. The stress at a point $\mathbf{r}$ influences that at another point $\mathbf{r}'$ through the function $g_{ik}^0(\mathbf{r}, \mathbf{r}')$ which weights the influence. Moreover, the role of the reference system enters through $g_{ik}^0(\mathbf{r}, \mathbf{r}')$ in such a way that suitable choices for the reference elasticity can improve the bounds without referring to detailed structural information. Techniques for choosing suitable reference systems will be described in a subsequent section.

All these advantages are based on the presumption that eqn. (59) together with the boundary condition (eqn. (60)) can be solved for g_{mi}^0 for a chosen reference system which has been supposed to be of bounded domain. However, an analytical solution of the Green's tensor function for systems of bounded domain of any shape is not possible at this time. Even for an unbounded domain, the Green's tensor function can be found in analytical form only for certain symmetries of C_{ijkl}^0, such as isotropic symmetry[16] and transverse isotropic symmetry.[18-20] However, analytical forms of the Green's tensor function in Fourier space can always be obtained for the case of unbounded domain.[21]

Fortunately, we are not concerned with boundary value problems. Our major concern is to find the averaged behaviour of a heterogeneous system. It is thus reasonable to extend the domain of the variational problem to infinity under the assumption that the system is of statistical homogeneity.

C.   Extension to the Unbounded Domain

The extension of the domain to infinity calls for proper normalisation since the total strain energy of an unbounded region is a divergent quantity. Thus for a system of unbounded domain, the average strain energy per unit volume, rather than the total energy, becomes the quantity of interest. It is straightforward to extend the boundary of Theorem 1 to infinity and simultaneously normalise the objective to unit volume. The resulting theorem reads:

Theorem 3: The normalised objective

$$\langle U_h \rangle = \langle U^0 \rangle - \tfrac{1}{2}\langle p_{ij}H_{ijkl}p_{kl} \rangle + \langle p_{ij}\varepsilon_{ij}^0 \rangle - \tfrac{1}{2}\langle u_i'p_{ij,j} \rangle \tag{72}$$

subject to the subsidiary conditions (eqns. (46), (48)) and

$$u_i'(|\mathbf{r}| \to \infty) = 0 \tag{73}$$

is stationary when eqn. (47) is satisfied. The equilibrium strain energy per unit volume of the actual system is given by $\langle U_h^s \rangle$ (the stationary value of $\langle U_h \rangle$) which satisfies the inequalities:

$$\langle U_h^s \rangle \geq \langle U_h \rangle \qquad \text{when } R_{ijkl} \text{ is positive definite} \tag{74}$$

while

$$\langle U_h^s \rangle \leq \langle U_h \rangle \qquad \text{when } R_{ijkl} \text{ is negative definite} \tag{75}$$

for $\langle U_h \rangle$ corresponding to admissible p_{ij} and ε_{ij}' fields. The notation $\langle \ \ \rangle$ stands for the operation of volume averaging, viz.,

$$\langle f \rangle = \lim_{V \to \infty} \frac{1}{V} \int f \, \mathrm{d}V \tag{76}$$

Following the same strategy for the reformulation of Theorem 2, we introduce the Green's tensor function corresponding to the reference elasticity for infinite domain, $g_{mi}^0(\mathbf{r})$, through the following equation and the boundary condition:

$$C_{ijkl}^0 \frac{\partial^2}{\partial x_l \partial x_j} g_{mi}^0(\mathbf{r}) + \delta_{mk}\delta(\mathbf{r}) = 0 \tag{77}$$

$$g_{mi}^0(|\mathbf{r}| \to \infty) = 0 \tag{78}$$

From the divergence theorem and the boundary condition on u_i' (eqn. (73)) and $g_{mi}^0(\mathbf{r})$ (eqn. (78)), we obtain the following integral relation:

$$\int g_{mi}^0(\mathbf{r} - \mathbf{r}')C_{ijkl}^0 u_{k,lj}'(\mathbf{r}') \, \mathrm{d}V' = \int u_k'(\mathbf{r}')C_{ijkl}^0 g_{mi,lj}^0(\mathbf{r} - \mathbf{r}') \, \mathrm{d}V' \tag{79}$$

which, when taking into account eqns. (46), (48) and (77), reduces to the following expression for $u_m'(\mathbf{r})$ in terms of $p_{ij,j}(\mathbf{r})$:

$$u_m'(\mathbf{r}) = \int g_{mi}^0(\mathbf{r} - \mathbf{r}')p_{ij,j}(\mathbf{r}') \, \mathrm{d}V' \tag{80}$$

With the help of eqn. (80), the last term in the objective (72) can then be written as

$$\langle u_i'p_{ij,j} \rangle = \int g_{ik}^0(\mathbf{r})\langle p_{ij,j}(\mathbf{r}')p_{kl,l}(\mathbf{r}' - \mathbf{r}) \rangle \, \mathrm{d}V \tag{81}$$

The tensor quantity $\langle p_{ij,j}(\mathbf{r}')p_{kl,l}(\mathbf{r}' - \mathbf{r})\rangle$ represents the two point correlation of the $p_{ij,j}$ field, which is a function of the separation distance $\mathbf{r}$ only for systems presumed to be of statistical homogeneity. Since the p_{ij} field will be used as the variation variable, it is more convenient to express this quantity in terms of the correlation of the p_{ij} field. This is developed in Appendix 3 with the following results:

$$\langle p_{ij,j}(\mathbf{r}')p_{kl,l}(\mathbf{r}' - \mathbf{r})\rangle = -\frac{\partial}{\partial x_j}\frac{\partial}{\partial x_l}\langle p_{ij}(\mathbf{r}')p_{kl}(\mathbf{r}' - \mathbf{r})\rangle \qquad (82)$$

Substituting eqn. (82) into eqn. (81) yields the following expression for the quantity $\langle u_i' p_{ij,j}\rangle$:

$$\langle u_i' p_{ij,j}\rangle = -\int g_{ik}^0(\mathbf{r})\frac{\partial^2}{\partial x_j \partial x_l}\langle p_{ij}(\mathbf{r}')p_{kl}(\mathbf{r}' - \mathbf{r})\rangle\,\mathrm{d}V \qquad (83)$$

Discontinuity in the derivative of the Green's tensor function, $g_{ik}^0(\mathbf{r})$, can cause problems with the use of the divergence theorem to eliminate the differential operator on the correlation function in the above expression. This difficulty can be circumvented by considering this integral in the Fourier space. The Fourier transform of a function and its inverse relation in three dimension space are defined as:

$$F[f(\mathbf{r})] = \bar{f}(\mathbf{k}) = \int f(\mathbf{r})\exp{(i\mathbf{k}\cdot\mathbf{r})}\,\mathrm{d}V \qquad (84)$$

$$F^{-1}[\bar{f}(\mathbf{k})] = f(\mathbf{r}) = \frac{1}{8\pi^3}\int \bar{f}(\mathbf{k})\exp{(-i\mathbf{k}\cdot\mathbf{r})}\,\mathrm{d}V_k \qquad (85)$$

where $\bar{f}(\mathbf{k})$ is the Fourier transform of $f(\mathbf{r})$; $\mathrm{d}V_k$ denotes the differential volume element in the Fourier (or $\mathbf{k}$) space. The integrations are taken over the whole spaces.

In order to facilitate the Fourier transformation, it is convenient to separate $\langle p_{ij}(\mathbf{r}')p_{kl}(\mathbf{r}' - \mathbf{r})\rangle$ into two parts, i.e.,

$$\langle p_{ij}(\mathbf{r}')p_{kl}(\mathbf{r}' - \mathbf{r})\rangle = P_{ijkl}(\mathbf{r}) + \langle p_{ij}\rangle\langle p_{kl}\rangle \qquad (86)$$

where

$$P_{ijkl}(\mathbf{r}) = \langle p_{ij}'(\mathbf{r}')p_{kl}'(\mathbf{r}' - \mathbf{r})\rangle \qquad (87)$$

$$p_{ij}'(\mathbf{r}) = p_{ij}'(\mathbf{r}) - \langle p_{ij}\rangle \qquad (88)$$

and $\langle p_{ij}\rangle$ is the volume average of the stress polarisation field. The function $P_{ijkl}(\mathbf{r})$ is expected to go to zero as $\mathbf{r}$ approaches infinity because of the lack of correlation for large $\mathbf{r}$. Consequently, this function is expected to possess a Fourier transform. It is obvious that the constant part of

$\langle p_{ij}(\mathbf{r}')p_{kl}(\mathbf{r}' - \mathbf{r})\rangle$, viz., $\langle p_{ij}\rangle\langle p_{kl}\rangle$, does not contribute to $\langle u_i'p_{ij,j}\rangle$. Accordingly, eqn. (83) may be written as:

$$\langle u_i'p_{ij,j}\rangle = -\int g_{ik}^0(\mathbf{r})P_{ijkl,jl}(\mathbf{r})\,\mathrm{d}V \tag{89}$$

The Fourier transform of $P_{ijkl,jl}(\mathbf{r})$ is related to the Fourier transform of $P_{ijkl}(\mathbf{r})$ by:

$$F[P_{ijkl,jl}(\mathbf{r})] = -k_j k_l \bar{P}_{ijkl}(\mathbf{k}) \tag{90}$$

where $\bar{P}_{ijkl}(\mathbf{k})$ denotes the Fourier transform of $P_{ijkl}(\mathbf{r})$. From the Parseval relations[22] and eqn. (90), $\langle u_i'p_{ij,j}\rangle$ given by eqn. (89) can also be equated to:

$$\langle u_i'p_{ij,j}\rangle = \left(\frac{1}{8\pi^3}\right)\int \bar{g}_{ik}^0(\mathbf{k})k_j k_l \bar{P}_{ijkl}^*(\mathbf{k})\,\mathrm{d}V_k \tag{91}$$

where $\bar{P}_{ijkl}^*(\mathbf{k})$ denotes the complex conjugate of $\bar{P}_{ijkl}(\mathbf{k})$. It can be readily shown that $\bar{P}_{ijkl}^* = \bar{P}_{ijkl}$ if $P_{ijkl}(\mathbf{r}) = P_{ijkl}(-\mathbf{r})$. Note also the symmetry properties for P_{ijkl}:

$$P_{ijkl} = P_{jikl} = P_{ijlk} \tag{92}$$

according to its definition. And if $P_{ijkl}(\mathbf{r}) = P_{ijkl}(-\mathbf{r})$, it has the additional symmetry property:

$$P_{ijkl} = P_{klij} \tag{93}$$

These symmetry properties are also valid for $\bar{P}_{ijkl}$.

In view of these symmetry properties, eqn. (91) can be written as:

$$\langle u_i'p_{ij,j}\rangle = -\left(\frac{1}{8\pi^3}\right)\int \bar{G}_{ijkl}^0(\mathbf{k})\bar{P}_{ijkl}^*(\mathbf{k})\,\mathrm{d}V_k \tag{94}$$

by defining a function $\bar{G}_{ijkl}^0(\mathbf{k})$ as:

$$\bar{G}_{ijkl}^0(\mathbf{k}) = -\tfrac{1}{4}[\bar{g}_{ik}^0(\mathbf{k})k_j k_l + \bar{g}_{jk}^0(\mathbf{k})k_i k_l + \bar{g}_{il}^0(\mathbf{k})k_j k_k + \bar{g}_{jl}^0(\mathbf{k})k_i k_k] \tag{95}$$

The inverse transformation of $\bar{G}_{ijkl}^0(\mathbf{k})$ defines $G_{ijkl}^0(\mathbf{r})$:

$$G_{ijkl}^0(\mathbf{r}) = \left(\frac{1}{2\pi}\right)^3 \int \bar{G}_{ijkl}^0(\mathbf{k})\exp(-i\mathbf{r}\cdot\mathbf{k})\,\mathrm{d}V_k \tag{96}$$

Consequently, $\langle u_i'p_{ij,j}\rangle$ can be expressed in terms of integration over the real space:

$$\langle u_i'p_{ij,j}\rangle = -\int G_{ijkl}^0(\mathbf{r})P_{ijkl}(\mathbf{r})\,\mathrm{d}V$$
$$= -\int G_{ijkl}^0(\mathbf{r})\langle p_{ij}'(\mathbf{r}')p_{kl}'(\mathbf{r}' - \mathbf{r})\rangle\,\mathrm{d}V \tag{97}$$

Similar to eqn. (65), $g_{ij}^0(\mathbf{r})$ for unbounded domain has the symmetry property:

$$g_{ij}^0(\mathbf{r}) = g_{ji}^0(-\mathbf{r}) \tag{98}$$

From its definition (eqn. (77)) it can readily be shown that

$$g_{ij}^0(\mathbf{r}) = g_{ij}^0(-\mathbf{r}) \tag{99}$$

Consequently

$$g_{ij}^0(\mathbf{r}) = g_{ji}^0(\mathbf{r}) \tag{100}$$

From these symmetry properties and eqn. (95), it can be shown that $G_{ijkl}^0(\mathbf{r})$, as well as $\bar{G}_{ijkl}^0(\mathbf{k})$, possesses the following symmetry properties:

$$G_{ijkl}^0 = G_{jikl}^0 = G_{klij}^0 \tag{101}$$

$$G_{ijkl}^0(\mathbf{r}) = G_{ijkl}^0(-\mathbf{r}) \tag{102}$$

Note that once the reference elasticity is specified, both $g_{ij}^0(\mathbf{r})$ and $G_{ijkl}^0(\mathbf{r})$ can be determined from eqns. (77), (78), and (95).

Substitution of eqn. (97) into (72) leads to the counterpart of Theorem 2 for unbounded domain:

Theorem 4: The normalised objective

$$\langle U_p \rangle = \langle U^0 \rangle - \tfrac{1}{2}\langle p_{ij}H_{ijkl}p_{kl}\rangle + \langle p_{ij}\varepsilon_{ij}^0\rangle$$
$$+ \tfrac{1}{2}\int G_{ijkl}^0(\mathbf{r})\langle p_{ij}'(\mathbf{r}')p_{kl}'(\mathbf{r}'-\mathbf{r})\rangle\,\mathrm{d}V \tag{103}$$

is stationary at

$$H_{ijkl}p_{kl} = \varepsilon_{ij}^0 + \int G_{ijkl}^0(\mathbf{r}-\mathbf{r}')p_{kl}'(\mathbf{r}')\,\mathrm{d}V' \tag{104}$$

and the equilibrium strain energy per unit volume, corresponding to the stationary value of $\langle U_p \rangle$, is given by

$$\langle U_p^s \rangle = \langle U^0 \rangle + \tfrac{1}{2}\langle p_{ij}^s\varepsilon_{ij}^0\rangle \tag{105}$$

where p_{ij}^s denotes the stationary solution of p_{ij} field, which satisfies eqn. (104). $\langle U_p^s \rangle$ also satisfies the inequalities:

$$\langle U_p^s \rangle \geq \langle U_p \rangle \qquad \text{when } R_{ijkl} \text{ is positive definite} \tag{106}$$

while

$$\langle U_p^s \rangle \leq \langle U_p \rangle \qquad \text{when } R_{ijkl} \text{ is negative definite} \tag{107}$$

Having arrived at this fundamental theorem for the variational treatments, we are in a position to construct the upper and lower bounds on the equilibrium strain energy of heterogeneous systems. Common procedures of finding bounds involve the proposal of a trial function for the approximation of the stress (or strain) field and the evaluation of the correspondent objective integral. The amount of structural information needed for this evaluation depends upon the structure of the proposed trial function. This implies that the choice of the trial function is by no means arbitrary; it should be so chosen that the correspondent objective potential can be evaluated with the information available, or accessible.

Theorem 4 provides considerable flexibility since the reference elasticity enters the objective as adjustable parameters. It is anticipated that by adjusting the moduli of the reference elasticity, better bounds can be generated. However, the exact role that the reference system plays cannot be fully appreciated before the behaviour of the function $G^0_{ijkl}(\mathbf{r})$ is understood. This tensor function may be viewed as a structure factor which weights the stress correlation function.

D. Properties of the Structure Factor Function

The Green's tensor function for a system of linear elasticity, C^0_{ijkl}, and of unbounded domain is defined as the solution of eqn. (77) with the boundary condition (eqn. (78)). The most straightforward and powerful method for solving this type of equation is the Fourier transformation, which transforms the differential equation into an algebraic equation. The solution can be immediately expressed in the Fourier space. However, the inverse transformation in terms of known functions is not always possible.

The Fourier transformation of eqn. (77) gives the following algebraic equation:

$$C^0_{ijkl}k_l k_j \bar{g}^0_{mi}(\mathbf{k}) = \delta_{mk} \tag{108}$$

Denote:

$$D_{ik} = C^0_{ijkl}k_l k_j \tag{109}$$

$$D_{ik}D^{-1}_{kn} = \delta_{in} \tag{110}$$

The Fourier transform of the Green's tensor function, $\bar{g}_{mn}(\mathbf{k})$, is obtained as:

$$\bar{g}^0_{mn}(\mathbf{k}) = D^{-1}_{mn} \tag{111}$$

Accordingly, the Green's tensor function, $g^0_{mn}(\mathbf{r})$, can be found by taking the inverse transformation of eqn. (111):

$$g^0_{mn}(\mathbf{r}) = \left(\frac{1}{2\pi}\right)^3 \int D^{-1}_{mn} \exp(-i\mathbf{r}\cdot\mathbf{k})\,dV_k \tag{112}$$

As mentioned earlier, it is possible to express $g^0_{mn}(\mathbf{r})$ in terms of known functions only for a few special cases, among which isotropic media yield the simplest results. In order to explore the behaviour of the function, $G_{ijkl}^0(\mathbf{r})$, we shall first consider the isotropic case. We shall then establish some general properties for $G_{ijkl}^0(\mathbf{r})$ and its role in the construction of bounds.

Isotropic Reference Media
If the reference elasticity is chosen to be isotropic, C_{ijkl}^0 can be expressed as in:[8]

$$C_{ijkl}^0 = \lambda^0 \delta_{ij}\delta_{kl} + 2\mu^0 I_{ijkl} \tag{113}$$

where λ^0 is known as the Lamé constant and μ^0 is the shear modulus (modulus of rigidity). Substituting eqn. (113) into eqn. (109) and taking the inverse of D_{ik} reveals, from eqn. (111), that:

$$g_{mn}^0(\mathbf{k}) = \frac{1}{\mu^0}\left(\frac{\delta_{mn}}{k^2} - \frac{\lambda^0 + \mu^0}{\lambda^0 + 2\mu^0}\frac{k_m k_n}{k^4}\right) \tag{114}$$

where

$$k^2 = k_i k_i \tag{115}$$

The inverse transformation of eqn. (114) is readily obtained for this case (see Appendix 4 for details). The result is

$$g_{mn}^0(\mathbf{r}) = \frac{1}{8\pi\mu^0}\frac{\lambda^0 + 3\mu^0}{\lambda^0 + 2\mu^0}\frac{\delta_{mn}}{r} + \frac{\lambda^0 + \mu^0}{\lambda^0 + 2\mu^0}\frac{x_m x_n}{r^3} \tag{116}$$

where

$$r^2 = x_i x_i \tag{117}$$

It is clear that the boundary condition (eqn. (78)) is satisfied by the Green's tensor function $g_{mn}^0(\mathbf{r})$ given by eqn. (116).

From the definition of $G_{ijkl}^0(\mathbf{r})$, viz., eqns. (95) and (96), it is shown in Appendix 4 that:

$$\bar{G}_{mnij}^0(\mathbf{k}) = \frac{\lambda^0 + \mu^0}{\mu^0(\lambda^0 + 2\mu^0)}\frac{k_m k_n k_i k_j}{k^4} - \frac{1}{4\mu^0 k^2}$$
$$\times (\delta_{mi}k_j k_n + \delta_{ni}k_j k_m + \delta_{mj}k_n k_i + \delta_{nj}k_i k_m) \tag{118}$$

$$G_{mnij}^0(\mathbf{r}) = \frac{\lambda^0 + \mu^0}{8\pi\mu^0(\lambda^0 + 2\mu^0)}\frac{1}{r^3}\left[\frac{15x_m x_n x_i x_j}{r^4} - 3\left(\delta_{ij}\frac{x_m x_n}{r^2} + \delta_{mn}\frac{x_i x_j}{r^2}\right) + \delta_{ij}\delta_{mn}\right.$$
$$\left. - \frac{3}{2}\frac{(3\lambda^0 + 4\mu^0)}{\lambda^0 + \mu^0}\left(\delta_{mi}\frac{x_n x_j}{r^2} + \delta_{ni}\frac{x_m x_j}{r^2} + \delta_{mj}\frac{x_n x_i}{r^2} + \delta_{nj}\frac{x_m x_i}{r^2}\right)\right.$$
$$\left. + \frac{2\lambda^0 + 3\mu^0}{\lambda^0 + \mu^0}2I_{mnij}\right] \tag{119}$$

A close examination of eqn. (119) reveals that $G^0_{mnij}(\mathbf{r})$ is a function of high singularity at the origin and that the integration of $G^0_{mnij}(\mathbf{r})$ over all the spacial directions at fixed $\mathbf{r} \neq 0$ vanishes. Thus a directionally averaged $G^0_{mnij}(\mathbf{r})$ behaves like a delta function with the infinity peak at the origin. In order to determine the magnitude of this delta function, we first denote the directionally averaged $G^0_{mnij}(\mathbf{r})$ by $F^0_{mnij}(r)$:

$$F^0_{mnij}(r) = \left(\frac{1}{4\pi}\right) \int_0^{2\pi} \int_0^{\pi} G^0_{mnij}(\mathbf{r}) \sin \vartheta \, d\vartheta \, d\phi \qquad (120)$$

where ϑ and ϕ are defined as

$$x_1 = r \cos \phi \sin \vartheta$$
$$x_2 = r \sin \phi \sin \vartheta \qquad (121)$$
$$x_3 = r \cos \vartheta$$

It is clear that $F^0_{mnji}(r)$ will not be a function in the usual sense. Accordingly, we must proceed in the framework of generalised functions by considering, instead, the following integral:

$$\int G^0_{mnij}(\mathbf{r}) f(r) \, dV$$

where $f(r)$ is assumed to be a function of $r = |\mathbf{r}|$ only. With the definition (eqn. (120)) this integral can be written as

$$\int G^0_{mnij}(\mathbf{r}) f(r) \, dV = \int F^0_{mnij}(r) f(r) \, dV \qquad (122)$$

For convenience, $f(r)$ will be treated as a regular function such that its Fourier transform exists. Because $f(r)$ is a radial function (a function of r only), it immediately follows that its Fourier transform is also a radial function such that $\bar{f}(\mathbf{k}) = \bar{f}(k)$.[22] It is also obvious that $f = f^*$ and $\bar{f} = \bar{f}^*$ where '*' denotes the complex conjugate. According to the Parseval's relations[22] and in view of the above properties associated with radial functions, the integral in eqn. (122) is equivalent to

$$\int F^0_{mnij}(r) f(r) \, dV = \left(\frac{1}{2\pi}\right)^3 \int \bar{G}^0_{mnij}(\mathbf{k}) \bar{f}(k) \, dV_k \qquad (123)$$

Introducing the polar coordinates in the Fourier space:

$$k_1 = k \cos \psi \sin \beta = k\alpha_1$$
$$k_2 = k \sin \psi \sin \beta = k\alpha_2 \qquad (124)$$
$$k_3 = k \cos \beta = k\alpha_3$$

and substituting eqn. (118) into eqn. (123) yields:

$$
\int F^0_{mnij}(r)f(r) = \left(\frac{1}{2\pi}\right)^3 \int_0^\infty \bar{f}(k)k^2 \, dk \left\{ \int_0^{2\pi} \int_0^\pi \left[\frac{\lambda^0 + \mu^0}{\mu^0(\lambda^0 + 2\mu^0)} \Big|_{\alpha_m \alpha_n \alpha_i \alpha_j} \right. \right.
$$
$$
\left. \left. - \frac{1}{4\mu^0}(\delta_{mi}\alpha_j\alpha_n + \delta_{ni}\alpha_j\alpha_m + \delta_{mj}\alpha_n\alpha_i + \delta_{nj}\alpha_m\alpha_i) \right] \sin\beta \, d\beta \, d\psi \right\}
$$

(125)

Note that the integration over the angles can readily be obtained by using the following relations:

$$
\int_0^{2\pi} \int_0^\pi \alpha_m \alpha_n \alpha_i \alpha_j \sin\beta \, d\beta \, d\psi = \frac{4\pi}{15}(\delta_{mn}\delta_{ij} + 2I_{mnij})
$$

(126)

$$
\int_0^{2\pi} \int_0^\pi \alpha_i \alpha_j \sin\beta \, d\beta \, d\psi = \frac{4\pi}{3} \delta_{ij}
$$

(127)

With the help of these relations, eqn. (125) can be written as

$$
\int F^0_{mnij}(r)f(r) \, dV = E^0_{mnij}\left(\frac{1}{2\pi}\right)^3 \int \bar{f}(k) \, dV_k
$$

(128)

where

$$
E^0_{mnij} = \frac{\lambda^0 + \mu^0}{15\mu^0(\lambda^0 + 2\mu^0)} \delta_{mn}\delta_{ij} - \frac{3\lambda^0 + 8\mu^0}{15\mu^0(\lambda^0 + 2\mu^0)} I_{mnij}
$$

(129)

From the definition of the Fourier inverse transformation, viz., eqn. (85), it is readily seen that:

$$
f(0) = \left(\frac{1}{2\pi}\right)^3 \int \bar{f}(\mathbf{k}) \, dV_k
$$

(130)

By eqn. (130), eqn. (128) can be written as:

$$
\int F^0_{mnij}(r)f(r) \, dV = E^0_{mnij}f(0)
$$

(131)

This implies that:

$$
F^0_{mnij}(r) = E^0_{mnij}\delta(\mathbf{r})
$$

(132)

since, by definition,

$$
\int f(r)\delta(\mathbf{r}) \, dV = f(0)
$$

(133)

This confirms that $F^0_{mnij}(r)$ indeed behaves like a delta function of strength

E^0_{mnij}. Equivalently, the volume integral of a function of spherical symmetry (of r dependence only) weighted by $G^0_{mnij}(\mathbf{r})$ depends only on the value of the function at $r = 0$, i.e.,

$$\int G^0_{mnij}(\mathbf{r})f(r)\,\mathrm{d}V = E^0_{mnij}.f(0) \qquad (134)$$

This striking property of G^0_{mnij} provides considerable insight into the construction of bounds. As noted previously, the evaluation of $\langle U_p \rangle$, defined in eqn. (103), must be carried out for every trial function selected to approximate the p_{ij} field. This evaluation, in general, demands information concerning the two point correlation function $P_{ijkl}(\mathbf{r}) = \langle p'_{ij}(\mathbf{r}')p'_{kl}(\mathbf{r}' - \mathbf{r})\rangle$, which appears as the integrand with weighting factor $G^0_{mnij}(\mathbf{r})$ in the last integral of eqn. (103). However, if the system of interest is known to be statistically isotropic such that $P_{ijkl}(\mathbf{r})$ can be made spherically symmetric by properly selecting the trial p_{ij} field, the amount of information needed to evaluate the said integral can be reduced to $P_{ijkl}(0)$. This is, of course, nothing more than the volume average of $p'_{ij}p'_{kl}$. Under this circumstance, it is expected that detailed correlation information is no longer necessary for constructing better bounds.

It will be shown later that $G^0_{mnij}(\mathbf{r})$, corresponding to reference elasticity of symmetries other than isotropy, still preserves this important property. It should also be noted that this property was derived by taking into account the fact that the Fourier transformation of a radial function preserves the radial symmetry in the Fourier space. Since a function of ellipsoidal symmetry can be treated as a radial function if proper coordinate transformations are made, we shall consider next the general situation when C^0_{ijkl} is of arbitrary symmetry and $f(\mathbf{r})$ is of ellipsoidal symmetry. This general consideration certainly contains the above treatment as a special case.

General Consideration

It was suggested in the analysis for isotropic reference media that the integral relation

$$\int G^0_{mnij}(\mathbf{r})f(\mathbf{r})\,\mathrm{d}V = E^0_{mnij}.f(0) \qquad (135)$$

(with E^0_{mnij} taken to be a constant tensor) can always be produced if $f(\mathbf{r})$ possesses ellipsoidal symmetry. We shall now consider the case in which $f(\mathbf{r})$ is of general ellipsoidal symmetry, namely, $f(\mathbf{r})$ is assumed to be a function of ρ only, where ρ is given by the following relation:

$$\rho^2 = (x_1/a_1)^2 + (x_2/a_2)^2 + (x_3/a_3)^2 \qquad (136)$$

Consequently, the Fourier transform of $f(\mathbf{r})$ is also of ellipsoidal symmetry.[23] It can readily be shown that $\bar{f}(\mathbf{k})$ is now a function of s only with s defined as:

$$s^2 = (a_1 k_1)^2 + (a_2 k_2)^2 + (a_3 k_3)^2 \qquad (137)$$

The integral on the left-hand side of eqn. (135) can be replaced by an integral in the Fourier space via Parseval's relations to give

$$\int G^0_{mnij}(\mathbf{r}) f(\mathbf{r})\, dV = \left(\frac{1}{2\pi}\right)^3 \int \bar{G}^0_{mnij}(\mathbf{k}) \bar{f}(\mathbf{k})\, dV_k \qquad (138)$$

In view of the symmetry assigned to $f(\mathbf{r})$, it is reasonable to make the following coordinate transformation in the Fourier space:

$$k_1 = (s/a_1) \cos \varphi \cdot \sin \eta = (s/a_1)\gamma_1 = s\zeta_1$$

$$k_2 = (s/a_2) \sin \varphi \, \sin \eta = (s/a_2)\gamma_2 = s\zeta_2 \qquad (139)$$

$$k_3 = (s/a_3) \cos \eta \qquad\quad = (s/a_3)\gamma_3 = s\zeta_3$$

The Jacobian associated with this transformation can easily be found to be:

$$\frac{\partial(k_1, k_2, k_3)}{\partial(s, \varphi, \eta)} = \frac{s^2 \sin \eta}{a_1 a_2 a_3} \qquad (140)$$

With this transformation, $\bar{f}(\mathbf{k})$ can be expressed in terms of s only. $\bar{G}^0_{mnij}(\mathbf{k})$ can be derived by combining eqns. (95), (109), (110) and (111). In terms of the new coordinates, eqn. (109) can be written as:

$$D_{ik} = s^2 C^0_{ijkl}\zeta_j\zeta_l = B_{ik}s^2 \qquad (141)$$

where

$$B_{ik} = C^0_{ijkl}\zeta_l\zeta_j \qquad (142)$$

The term B_{ik} is clearly a function of φ and η only. Denote the inverse of B_{ik} by B_{ik}^{-1}. From eqns. (110) and (111), it is obvious that

$$\bar{g}^0_{mn} = B_{mn}^{-1}/s^2 \qquad (143)$$

Equation (95) can then be written as

$$\bar{G}^0_{ijkl} = -\tfrac{1}{4}(B_{ik}^{-1}\zeta_j\zeta_l + B_{jk}^{-1}\zeta_i\zeta_l + B_{il}^{-1}\zeta_j\zeta_k + B_{jl}^{-1}\zeta_i\zeta_k) \qquad (144)$$

Note that both B_{ik}^{-1} and ζ_is are functions of φ and η only. Consequently, $\bar{G}^0_{ijkl}$ is also a function of φ and η only and can hence be denoted as:

$$\bar{G}^0_{ijkl} = \bar{G}^0_{ijkl}(\varphi, \eta) \qquad (145)$$

In terms of the new coordinates, eqn. (138) becomes:

$$\int G^0_{mnij}(\mathbf{r}) f(\mathbf{r}) \, dV = \left(\frac{1}{2\pi}\right)^3 \int_0^\pi \int_0^{2\pi} \int_0^\infty \bar{G}^0_{mnij}(\varphi, \eta) \bar{f}(s) \cdot \frac{s^2 \sin \eta}{a_1 a_2 a_3} \, ds \, d\varphi \, d\eta$$

(146)

Define E^0_{mnij} as

$$E^0_{mnij} = \left(\frac{1}{4\pi}\right) \int_0^\pi \int_0^{2\pi} \bar{G}^0_{mnij}(\varphi, \eta) \sin \eta \, d\varphi \, d\eta$$

(147)

Equation (146) can then be written as:

$$\int G^0_{mnij}(\mathbf{r}) f(\mathbf{r}) \, dV = E^0_{mnij} \left(\frac{1}{2\pi}\right)^3 \int_0^\infty 4\pi \bar{f}(s) \frac{s^2 \, ds}{a_1 a_2 a_3}$$

$$= E^0_{mnij} \left(\frac{1}{2\pi}\right)^3 \int \bar{f}(\mathbf{k}) \, dV_k$$

$$= E^0_{mnij} f(0)$$

(148)

Hence, the desired relationship (eqn. (135)) has been generated. These results lead to the following theorem:

Theorem 5: For a function $f(\mathbf{r})$ of ellipsoidal symmetry such that it is a function only of ρ (defined in eqn. (136)) the relationship (135) exists where E^0_{mnij} is a constant tensor given by eqn. (147).

E^0_{mnij} corresponding to a special case of ellipsoidal symmetry, viz., $a_1 = a_2 = 1$ and $a_3 = a$, is derived in Appendix 5. These preliminaries on the properties of the function $G^0_{mnij}(\mathbf{r})$ provide a basis for discussions in the next section concerning the construction of improved bounds.

IV. GENERAL BOUNDING FORMULATION

A. Preliminary Definitions

Before proceeding with the application of the new variational theorems to the development of a general bounding theory, it will be useful to introduce some definitions and preliminary statistical notations. First, a distinction is made between two space scales, viz., microscopic and macroscopic. The microscopic scale is comparable with the dimension of phase regions, which are large enough to be regarded as continua but are very small on the macroscopic scale in comparison to the size of the bulk specimen. In a heterogeneous system, the elasticity varies on the microscopic scale in a stochastic manner consistent with the microstructure. Since a detailed knowledge of the microstructure is not usually available, it is more useful to characterise the elasticity field by its various averages, such as volume average and multi-point correlations. In general, these averages are subject to fluctuations. It is, however, assumed that the size of the specimen is large enough so that the fluctuations are too small to detect. It is also assumed that the heterogeneous systems under consideration are statistically homogeneous in the usual sense of statistical mechanics such that the systems appear homogeneous, macroscopically. In virtue of these assumptions, an actual specimen of finite size can be represented by the proper normalisation of an infinite system characterised by the same homogeneous statistics. This assertion permits the use of Theorem 4 for the construction of bounds on the effective properties.

The volume average of some field property f, denoted by $\langle f \rangle$, is defined as:

$$\langle f \rangle = \langle f(\mathbf{r}) \rangle = \frac{1}{V} \int f(\mathbf{r}) \, dV \qquad (149)$$

where the volume should be large enough to smooth fluctuations. Another important statistical average is the two point correlation between properties f_1 and f_2 at two points separated by distance $\mathbf{r}$, viz.,

$$\langle f_1(\mathbf{r}')f_2(\mathbf{r}' - \mathbf{r}) \rangle = \frac{1}{V'} \int f_1(\mathbf{r}')f_2(\mathbf{r}' - \mathbf{r}) \, dV' \qquad (150)$$

where the integration is taken over the $\mathbf{r}'$ space and again V' should be large enough to smooth fluctuation. Note that $\langle f_1(\mathbf{r}')f_2(\mathbf{r}' - \mathbf{r}) \rangle$ is a function of the separation distance $\mathbf{r}$ only. Similarly, three point correlation, four point correlation, etc., can be defined.

In many systems, statistical symmetries are observed or expected. For example, polycrystalline aggregates resulting from the unconstrained solidification of melts are likely to behave isotropically, macroscopically. The statistical correlation of these systems should be invariants under any rotation transformation of the coordinates. Statistical symmetries are also expected in many fibre-reinforced materials, in which the statistical correlations are likely to be invariant under the rotation transformation about the axial (fibre) direction. A system is of statistical isotropy if all its statistical correlations are invariants under any rotation transformation of the coordinates. Similarly, a system is of statistical transverse isotropy if all its statistical correlations are invariants under any rotation transformation about a specific axis. Clearly, there are other possible symmetries. The following general symmetry situation, concerning the two point correlation function, is of particular importance to the subsequent development of the current bounding theory.

A two point correlation function $\langle f_1(\mathbf{r}')f_2(\mathbf{r}' - \mathbf{r}) \rangle$ is of ellipsoidal symmetry if it can be written as a function of a scalar ρ only, where ρ is given by:

$$\rho = [(x_1/a_1)^2 + (x_2/a_2)^2 + (x_3/a_3)^2]^{1/2} \tag{151}$$

Under the special case when a_1, a_2, and a_3 are mutually equal, the correlation function is of spherical symmetry. It is obvious that the two point correlation function of a statistically isotropic system possesses spherical symmetry.

Another interesting special case corresponds to continuous fibrous materials, including fibre-reinforced materials, which are of statistical transverse isotropy. Continuous fibrous materials are identified as those whose phase regions can be generated by sweeping some plane geometry along an unique direction which is referred to as the axial direction. The plane perpendicular to the axial direction is referred to as the transverse plane. It is obvious that all the transverse planes are of the same geometrical structure. Let the axial direction be along the x_3 axis. Because of the equivalency of all the transverse planes, the statistical correlations of a continuous fibrous material are functions of x_1 and x_2 only. By further assuming statistical transverse isotropy, they are functions of scalar $x_1^2 + x_2^2$ only. In this case, the two point correlation function corresponds to the special case of ellipsoidal symmetry with $a_1 = a_2 = 1$ and $a_3 = \infty$.

Another limiting case with $a_1 = a_2 = \infty$ corresponds to the trivial situation of a sandwich type of composite. The above three special cases cover all the situations under which improved bounding techniques without detailed statistics have been considered thus far.[2, 3, 6, 35] Unfortunately, most of the materials of engineering interest do not fall into these special symmetry situations. Accordingly, it is useful to consider the more general case of ellipsoidal symmetry, which contains these situations as special cases. This symmetry will be used to construct an improved bounding theory of sufficient generality for important engineering applications.

It will be fruitful to clarify some concepts concerning statistical symmetry before proceeding with the development of the bounding theory. First, it should be noted that the statistical symmetry of a system does not necessarily have to be in consistency with the symmetry of its macroscopic effective elasticity. To illustrate this point, consider a statistical isotropic heterogeneous system composed of isotropic phases. Clearly, this system behaves isotropically, macroscopically. Now, let us replace one of the phase media by an anisotropic material perfectly aligned in space. The resulting system still preserves statistical isotropy, but certainly behaves anisotropically on the average. On the other hand, macroscopically isotropic systems do not necessarily possess statistical isotropy because preferred orientations of the anisotropic phases might happen to balance the effect of statistical anisotropy to result in an overall isotropy. Second, it should be noted that the statistical symmetry is, in general, more than a geometrical structural descriptor. For systems composed of isotropic phases, statistical symmetry is completely determined by the geometrical structure. However, if anisotropic phases are present, orientation distributions might also take part in the determination of statistical symmetry. Only under the situation when there is no correlation between the phase region geometry and the orientation, the statistical symmetry is again a pure geometrical descriptor.

B.   Application of the Refined Variational Principles

Variational principles based on an unspecified reference system were developed in the previous section. The final form is summarised in Theorem 4, which states:

For a system under load such that the average strain is given by ε_{ij}^0, the equilibrium strain energy per unit volume should be given by the stationary value of the normalised objective, $\langle U_p \rangle$, defined as:

$$\langle U_p \rangle = \tfrac{1}{2}\varepsilon_{ij}^0 C_{ijkl}^0 \varepsilon_{kl}^0 - \tfrac{1}{2}\langle p_{ij} H_{ijkl} p_{kl} \rangle + \langle p_{ij} \rangle \varepsilon_{ij}^0$$
$$+ \tfrac{1}{2} \int G_{ijkl}^0(\mathbf{r}) \langle p_{ij}'(\mathbf{r}') p_{kl}'(\mathbf{r}' - \mathbf{r}) \rangle \, dV \qquad (152)$$

where the stationary value of $\langle U_p \rangle$, denoted by $\langle U_p^s \rangle$, corresponds to a maximum when R_{ijkl} is positive definite, and to a minimum when R_{ijkl} is negative definite.

The effective elasticity C_{ijkl}^* of the heterogeneous system is defined through the following relation:

$$\langle U_p^s \rangle = \tfrac{1}{2}\varepsilon_{ij}^0 C_{ijkl}^* \varepsilon_{kl}^0 \qquad (153)$$

The task at hand is to construct bounds on the equilibrium strain energy, $\langle U_p^s \rangle$, and hence on the effective elasticity C_{ijkl}^* as well, based on this variational principle and the bounding procedures outlined in the Introduction. The closeness of the bounds will certainly depend on the amount of information provided. Since the statistical details of a heterogeneous system are usually not available, the subsequent consideration will be limited to bounds resulting from two levels of information, beyond which statistical details must be introduced. The structural information needed to perform volume average, such as volume fractions of various phases and orientation distributions (if the phases exhibit anisotropy), are grouped in the first level as the minimum information required for a reasonable estimate of the effective performance. The statistical symmetry provides the second level of information. The objective here is to find the best possible bounds corresponding to each level of information.

C.   The Best Possible Bounds for the First Level of Information

For a system in which the only available information is that needed to evaluate the volume average of the elastic field, the choice of trial p_{ij} fields will be quite restricted. This situation is a consequence of the necessity of evaluating the correspondent objective integral. The limitation that the extent of structural knowledge imposes on the functional form of feasible trial p_{ij} fields is reflected mainly through the last integral term in the objective given by eqn. (152). In general, the evaluation of this term demands correlation information unless p_{ij}^t is chosen to be a constant tensor. In that case, $p_{ij}^{\prime t}$ is identically zero and this integral term vanishes. Since statistical correlations of any form are presumably unavailable for the first level of information, the suitable trial fields should be of the form:

$$p_{ij}^t = \text{constant tensor} \qquad (154)$$

The correspondent objective can thus be written as:

$$\langle U_p^t \rangle = \tfrac{1}{2}\varepsilon_{ij}^0 C_{ijkl}^0 \varepsilon_{kl}^0 - \tfrac{1}{2}p_{ij}^t \langle H_{ijkl} \rangle p_{kl}^t + p_{ij}^t \varepsilon_{ij}^0 \qquad (155)$$

where p_{ij}^t is now taken to be a constant tensor.

The optimal choice of p_{ij}^t can be obtained by taking the derivative of $\langle U_p^t \rangle$ with respect to p_{ij}^t and setting the result equal to zero. This gives:

$$-\langle H_{ijkl} \rangle p_{kl}^t + \varepsilon_{ij}^0 = 0 \qquad (156)$$

Substituting eqn. (A6-15) into (A6-14) yields the objective corresponding to the optimal p_{ij}^t, which is given by:

$$\langle U_p^t \rangle = \tfrac{1}{2}\varepsilon_{ij}^0 (C_{ijkl}^0 + \langle H_{ijkl} \rangle^{-1})\varepsilon_{kl}^0 \qquad (157)$$

The reference elasticity C_{ijkl}^0 is as yet unspecified. To generate a lower bound, C_{ijkl}^0 must be chosen so that the R_{ijkl} field is always positive definite. A careful examination of eqn. (157) reveals that, when R_{ijkl} is kept positive definite, $\langle U_p^t \rangle$ decreases monotonically as C_{ijkl}^0 increases. This implies that the greatest lower bound corresponds to the lowest possible C_{ijkl}^0, viz., $C_{ijkl}^0 = 0$. From eqn. (157), it is apparent that this greatest lower bound is given by the Reuss average, viz.,

$$(\langle U_p^s \rangle)_{\text{glb}} = \tfrac{1}{2}\varepsilon_{ij}^0 \langle S_{ijkl}(\mathbf{r}) \rangle^{-1} \varepsilon_{kl}^0 \leq U_p^s \tag{158}$$

where $(\langle U_p^s \rangle)_{\text{glb}}$ represents the greatest lower bound (glb) on the equilibrium strain energy, $\langle U_p^s \rangle$.

Similarly, in order to generate an upper bound, C_{ijkl}^0 must be chosen so that the R_{ijkl} field is always negative definite. Again, an examination of eqn. (157) reveals that, when R_{ijkl} is kept negative definite, $\langle U_p^t \rangle$ decreases monotonically as C_{ijkl}^0 increases. This implies that the least upper bound (lub) corresponds to the limit when C_{ijkl}^0 approaches infinity. This least upper bound will be denoted by $(\langle U_p^s \rangle)_{\text{lub}}$, which can be evaluated through the following limiting procedure:

$$
\begin{aligned}
2(\langle U_p^s \rangle)_{\text{lub}} &= \lim_{C^0 \to \infty} \varepsilon_{ij}^0 (C_{ijkl}^0 + \langle H_{ijkl} \rangle^{-1}) \varepsilon_{kl}^0 \\
&= \varepsilon_{ij}^0 \lim_{C^0 \to \infty} [C_{ijkl}^0 - \langle S_{ijop}^0 (I_{opkl} - C_{opmn} S_{mnkl}^0)^{-1} \rangle^{-1}] \varepsilon_{kl}^0 \\
&= \varepsilon_{ij}^0 \lim_{C^0 \to \infty} [C_{ijkl}^0 - \langle S_{ijop}^0 (I_{opkl} + C_{opmn} S_{mnkl}^0 + O(\mathbf{S}^{0^2})) \rangle^{-1}] \varepsilon_{kl}^0 \\
&= \varepsilon_{ij}^0 \lim_{C^0 \to \infty} [C_{ijkl}^0 - (I_{ijop} - \langle C_{ijmn} \rangle S_{mnop}^0 + O(\mathbf{S}^{0^2})) C_{opkl}^0] \varepsilon_{kl}^0 \\
&= \varepsilon_{ij}^0 \lim_{S^0 \to 0} [\langle C_{ijkl} \rangle + O(\mathbf{S}^0)] \varepsilon_{kl}^0 \\
&= \varepsilon_{ij}^0 \langle C_{ijkl} \rangle \varepsilon_{kl}^0
\end{aligned}
$$

or

$$(\langle U_p^s \rangle)_{\text{lub}} = \tfrac{1}{2}\varepsilon_{ij}^0 \langle C_{ijkl}(\mathbf{r}) \rangle \varepsilon_{kl}^0 \geq \langle U_p^s \rangle \tag{159}$$

This is the Voigt average.

Note that eqns. (158) and (159) hold for an arbitrary infinitesimal strain tensor ε_{ij}^0. This implies that $\langle C_{ijkl} \rangle - C_{ijkl}^*$ and $C_{ijkl}^* - \langle S_{ijkl} \rangle^{-1}$ must always be positive semi-definite, where C_{ijkl}^* is the effective elasticity defined in eqn. (153). Thus, in this sense, C_{ijkl}^* is bounded from both above and below by the Voigt and Reuss averages, respectively.

In summary, for a system in which the only available information is that needed to evaluate volume averages, the trial stress field, p_{ij}^t, must be taken as a constant tensor in order not to involve inaccessible information. Consequently, the Reuss and Voigt averages are the greatest lower bound and the least upper bound, respectively, that can be obtained within the first level of information.

D. Improved Bounds for Systems of Statistical Ellipsoidal Symmetry

The Reuss and Voigt averages are the best possible bounds available within the first level of information. In order to improve these bounds, additional structural information is absolutely necessary. Additional information may be available in various forms. It is highly likely that not every piece of additional information will contribute to an improvement in the bounds, either because of the limitations of its form or contents, or because the formulation has not been cast into a form which can use the information. Rather than divert attention to a discussion of the nature and quality of existing experimental structural information, it will be more fruitful to develop a bounding formulation which incorporates a specific type of structural information, namely, the statistical symmetries. Features of the statistical symmetries will then be related to structural characteristics that may be determined or inferred from structural data.

For this purpose, attention is directed to a system which possesses ellipsoidal symmetry as far as the two point correlation function of its elastic field is concerned. It is also assumed that the information needed to evaluate volume averages is provided. However, detailed field correlations will be ignored. Again, the last integral term in eqn. (152) dominates the determination of the form of feasible trial p_{ij} fields. Under the current level of structural knowledge, we are no longer confined to the simple trial fields of constant p'_{ij}. Theorem 5 reveals that if $\langle p''^t_{ij}(\mathbf{r}')p''^t_{kl}(\mathbf{r}' - \mathbf{r})\rangle$ is taken to be of ellipsoidal symmetry, the last integral term in eqn. (152) can be evaluated without resorting to detailed correlation information.

Accordingly, the trial p_{ij} fields are selected in such a form that the p'_{ij} at a certain point in space is a function of its local elasticity field only, viz.,

$$p'_{ij}(\mathbf{r}) = \text{function of } C_{ijkl}(\mathbf{r}) \tag{160}$$

Physically, this $p'_{ij}(\mathbf{r})$ is a piecewise constant function for a multiphase system. It will be confirmed later that this is the best proposition for the current level of structural knowledge. A more realistic trial field will make it impossible to evaluate the correspondent objective without involving other (presumably unavailable) information.

Under the assumption that the two point correlation function of the elastic field is of ellipsoidal symmetry, it can be shown that the two point correlations of the proposed trial p_{ij} fields also possess the same symmetry, viz.,

$$\langle p''^t_{ij}(\mathbf{r}')p''^t_{kl}(\mathbf{r}' - \mathbf{r})\rangle = f(\rho) \tag{161}$$

where ρ is defined in eqn. (151) with a_1, a_2, and a_3 given. With the help of

Theorem 5, the last integral term in eqn. (152) can be written as

$$\tfrac{1}{2}\int G^0_{ijkl}(\mathbf{r})\langle p''_{ij}(\mathbf{r}')p''_{kl}(\mathbf{r}'-\mathbf{r})\rangle\,dV = \tfrac{1}{2}E^0_{ijkl}\langle p''_{ij}p''_{kl}\rangle \qquad (162)$$

where E^0_{ijkl} is a constant tensor given by eqn. (147). Note that $\langle p''_{ij}p''_{kl}\rangle$ is nothing more than the volume average of $p''_{ij}p''_{kl}$ field, which is presumably evaluatable. Substituting the proposed p^t_{ij} and eqn. (162) into eqn. (152) leads to the following expression for the correspondent objective $\langle U^t_p\rangle$:

$$\langle U^t_p\rangle = \tfrac{1}{2}\varepsilon^0_{ij}C^0_{ijkl}\varepsilon^0_{kl} - \tfrac{1}{2}\langle p^t_{ij}H_{ijkl}p^t_{kl}\rangle + \langle p^t_{ij}\rangle\varepsilon^0_{ij} + \tfrac{1}{2}\langle p''_{ij}E^0_{ijkl}p''_{kl}\rangle \quad (163)$$

The optimal p^t_{ij} field is then determined by equating the variation of $\langle U^t_p\rangle$ with respect to p^t_{ij} to zero. This gives the relation for the optimal p^t_{ij}:

$$H_{ijkl}p^t_{kl} = \varepsilon^0_{ij} + E^0_{ijkl}p''_{kl} \qquad (164)$$

From eqn. (164), $\langle U^t_p\rangle$, corresponding to the optimal p^t_{ij}, can be written as

$$\langle U^t_p\rangle = \tfrac{1}{2}\varepsilon^0_{ij}C^0_{ijkl}\varepsilon^0_{kl} + \tfrac{1}{2}\langle p^t_{ij}\rangle\varepsilon^0_{ij} \qquad (165)$$

where $\langle p^t_{ij}\rangle$ can be determined from eqn. (164) by simple algebraic manipulation which leads to the following relation:

$$\langle p^t_{ij}\rangle = (I_{ijmn} + \langle M_{ijop}\rangle E^0_{opmn})^{-1}\langle M_{mnkl}\rangle\varepsilon^0_{kl} = (\langle M_{ijkl}\rangle^{-1} + E^0_{ijkl})^{-1}\varepsilon^0_{kl}$$

$$(166)$$

where M_{ijkl} is defined as:

$$M_{ijkl} = (H_{ijkl} - E^0_{ijkl})^{-1} \qquad (167)$$

Combining eqns. (165) and (166), we can express $\langle U^t_p\rangle$ as:

$$\langle U^t_p\rangle = \tfrac{1}{2}\varepsilon^0_{ij}C^t_{ijkl}\varepsilon^0_{kl} \qquad (168)$$

with C^t_{ijkl} defined as:

$$C^t_{ijkl} = C^0_{ijkl} + (\langle M_{ijkl}\rangle^{-1} + E^0_{ijkl})^{-1} \qquad (169)$$

Equations (168) and (169) provide for the construction of improved bounds for the effective moduli of heterogeneous systems of statistical ellipsoidal symmetry. The average of a tensorial quantity, such as M_{ijkl} in eqn. (169), subject to a certain orientation distribution, is discussed in Appendix 6.

Whether eqn. (169) corresponds to a lower bound or to an upper bound depends on the choice for the reference elasticity C^0_{ijkl}. If C^0_{ijkl} is chosen such that the R_{ijkl} field is always positive definite, a lower bound is obtained. Alternatively, if the R_{ijkl} field is kept negative definite, an upper bound is obtained. A qualitative examination of eqn. (169) immediately reveals that C^t_{ijkl} increases as C^0_{ijkl} increases. This implies that for the case of lower bound, C^0_{ijkl} should be chosen as large as possible in order to obtain the best possible lower bound; while C^0_{ijkl} should be chosen as small as possible to obtain the best possible upper bound. It is interesting to note that the Reuss and Voigt averages can be generated by setting C^0_{ijkl} equal to zero and infinity, respectively, in eqn. (169). Consequently, other choices of C^0_{ijkl} will always lead to bounds inbetween the Reuss and Voigt limits.

It can be shown[26] that the bounding formula equation (169) includes various previous bounding treatments[1−6] as special cases. This formula can also be reduced to various existing micromechanical models as will be discussed in next section.

It was asserted that the piecewise constant p^t_{ij} is the best possible form for the trial fields under the current level of structural knowledge. This assertion can be confirmed through the following arguments. First, we relax the assumption of piecewise constancy on p^t_{ij}, and do not assign any specific functional form to the trial fields at the outset. However, it is necessary to demand that the p^t_{ij} be proposed such that the resulting two point correlation function is of ellipsoidal symmetry, because otherwise the correspondent objective will not be evaluatable. In consequence of this requirement, it immediately follows that the correspondent objective is again given by eqn. (163), which, upon optimisation, generates a piecewise constant field given by eqn. (164). It can thus be concluded that the piecewise constant p^t_{ij} is indeed the best possible form for the trial fields. Accordingly, the resulting optimal bounds given by eqn. (169) should be the best possible bounds under the current level of structural information, provided that the optimal reference elasticity is selected.

Clearly, the classical variational principles do not provide guidelines to assist in developing a proposition for better trial fields, nor do they reveal whether and what additional information is required. These are the serious limitations of the classical variational principles. Unfortunately, the variational principle proposed by Hashin and Shtrikman,[1] as summarised in Theorem 1, also fails to provide guidance for the choice of better trial fields. In this treatment, however, the application of Theorem 4 not only suggests the forms of the trial functions under various circumstances, but also identifies the best possible bounds consistent with the level of available information.

V.   REDUCTION TO MODELS

A.   Bounding Formulae as Models

The best possible bounds for various levels of structural information were developed in the previous section. These bounds are valid only if the information incorporated is known precisely. The primitive Reuss and Voigt bounds can be established from limited structural information, viz., the gross phase concentrations and orientation distributions. These structural descriptors are easily obtained by well-known techniques and may be controlled by adjusting processing conditions. Any improvement (i.e. tightening) of these bounds requires additional information concerning the nature of the statistical symmetry. Specifically, it was shown that improved bounds can only be rigorously constructed for the case of precise statistical ellipsoidal symmetry. Under the prevalent approaches toward experimental techniques and methods of data analysis, it is difficult to secure precise information on any structural statistics other than simple distributions. At best, only rough estimates of the statistical symmetry are currently available. Under these circumstances, the improved bounds should be interpreted as 'model' bounds dictated by the behaviour of a hypothetical system which may only approximately represent reality.

Indeed, the bound itself can be treated as a model of the system in the sense that it represents some hypothetical physical situation which mimics a real system. For example, the Reuss bound has been used extensively as a model to represent the behaviour of a dispersed phase in a continuous medium. This simple model is not satisfactory; it reveals little concerning structure–property relationships and hence can hardly serve to guide material improvements. The Reuss model actually gives the lower limit to the material performance. However, in principle, the material behaviour can approach the upper limit of the Voigt model. Improvement in the behaviour of a material from its lower performance limit toward the upper limit requires an understanding of the structural parameters which govern the intermediate behaviour of the material between these two limits. The empirical ξ factor in the Tsai–Halpin equation[24] was introduced to serve as such a parameter. However, the validity of this equation and the correlation of ξ to the aspect ratio of the dispersed phase have not been generally justified. Moreover, the simple algebraic form precludes applications to more complicated situations involving complex phase geometry or anisotropic phase orientations.

Fortunately, the improved bounding formulae, eqn. (169), introduce useful new parameters. The tensor quantity E^0_{ijkl} incorporates, via eqn. (147), the statistical symmetry which performs a role similar to the ξ factor of the Tsai–Halpin equation; the reference elasticity also enters the formulation as adjustable parameters. Accordingly, it is worthwhile to explore the status of these improved bounding formulae as explicit models for heterogeneous systems. For this purpose, it is instructive to re-examine the derivation of eqn. (169) from a more general point of view. Note that the reference elasticity parameters in the objective (eqn. (152)) determine whether the stationary solution corresponds to a maximum or a minimum. However, it can be rigorously shown that the choice of the reference elasticity does not change the stationary value of the objective (eqn. (152)). One approach to obtaining the stationary solution is to solve the integral equation (104) through the method of successive approximations. This approach will involve statistical correlations of all orders.[25] Because of the lack of such information, attention was restricted to the consideration of a piecewise constant p_{ij}^t field and a resultant correspondent objective, $\langle U_p^t \rangle$, given by eqn. (163). The subsequent optimisation of eqn. (163) led to the improved bounding formulae (eqn. (169)), which corresponds to the stationary value of eqn. (163). Note that the stationary value of eqn. (163) is now a function of the reference elasticity and, in general, differs from the equilibrium strain energy of the system, given by the stationary value of the original objective (eqn. (152)). However, it is expected that for a certain choice of the reference elasticity, these two stationary values will coincide. Clearly, based on this reference elasticity, eqn. (169) predicts the effective moduli of the heterogeneous system.

There should be physical significance associated with such a explicit choice of the reference elasticity. Note that the objective (eqn. (163)) corresponding to the piecewise constant p_{ij}^t does not contain a term characterising the nature of the correlation of the stress field, as does the original objective (eqn. (152)). However, both objectives result in the same stationary value for a certain choice of reference elasticity; this suggests that the implicit stress correlation can be accounted for by assigning such a reference elasticity to the objective (eqn. (163)). The role that the C_{ijkl}^0s play in eqn. (169) should thence be interpreted as a factor that takes into account the contribution of the stress correlation to the strain energy. Obviously, such an interpretation implies that the reference elasticity C_{ijkl}^0 should be, in general, a function of all the structural statistics—both known and unknown. Accordingly, because of this lack of information, the C_{ijkl}^0 will be treated as a constant tensor. This assertion introduces eqn. (169) as a model. Under most situations, it is reasonable to expect that this assertion is valid; the major contributions from such dominant structural factors as volume fractions, orientation distributions and statistical symmetry have already been isolated in the development of eqn. (169).

Even under this simplification of treating C_{ijkl}^0 as a constant tensor, the determination of its moduli for a given system remains a difficult problem. This is again attributed to the lack of information. Therefore for the practical use of eqn. (169) as a model, it is appropriate to specify the reference elasticity by physical judgements derived from qualitative observations. This situation demands a deeper understanding of the physical meaning of the model equation (169) and the associated parameters.

B. Physical Significance of the Statistical Symmetry Descriptors and the Reference Elasticity

Since the effective elasticity is, by definition, the measure of the strain energy density in a heterogeneous system under load, it is related to, and could be determined from, a knowledge of the stress (or strain) distribution within the system. In principle, such a knowledge is deducible from the solution of a set of governing equations consistent with the microstructure of the system. However, the microstructures of heterogeneous systems are so complicated and indeterminate that only partial and/or qualitative structural descriptions are usually available. As a consequence, the exact stress (or strain) distribution cannot be determined. In view of this limitation, the common thrust of various micromechanical models has centred on the assertion of some tractable stress (or strain) distribution that hopefully approximates the actual field.

The Reuss and Voigt averages correspond to the hypothetical physical states in which either the stress or the strain field is assumed constant. Such a simplification implies that all the material elements in a heterogeneous system are either uniformly stressed or uniformly strained regardless of the shapes, properties and locations of the elements. Accordingly, the resulting averaged properties depend only on such structural descriptors as volume fractions and orientation distributions. Clearly, the actual stress and strain distributions within a loaded heterogeneous medium must be quite complicated. Moreover, the exact values of the effective moduli depend on how the load is shared among the various phases in a manner intermediate between the extreme situations of the Reuss and Voigt models. It is reasonable to expect that under the same state of volume fractions and orientation distributions, one heterogeneous system may be stiffer than another ostensibly equivalent system if the stiffer phase shares a greater portion of the local load.

The load distribution depends partly on the microstructure of the heterogeneous system. For example, consider a composite, containing stiff inclusions embedded in a soft matrix, subject to an unidirectional stretch. The shapes of the inclusions play an important role in the determination of the elastic response; inclusions of elongated shapes along the loading direction will share more load than short inclusions. This implies that the aspect ratio of the inclusions, defined as the quotient of the axial length over the lateral diameter, may be the dominant geometrical descriptor in the determination of effective moduli. This geometrical descriptor enters the formulation through the statistical symmetry. The set of parameters $(a_1, a_2$ and $a_3)$ that characterise the correlation symmetry can be viewed, in an averaged sense, as related to the geometrical aspect ratios along various directions. The dependence of the effective moduli on these 'statistical aspect ratios' is illustrated by the special case of ellipsoidal symmetry with $a_1 = a_2 = 1$ and $a_3 = a$. If the '3' axis is taken as the axial direction, the corresponding aspect ratio is given by a. Clearly, this special statistical symmetry is closely associated with microstructures resulting from unidirectional processes. The value of a can vary from zero to infinity. Values of $a \to 0$ may be associated with a disc-like microstructure of the reinforcing phase while values for $a \to \infty$ correspond to continuous fibre reinforced systems. A value of $a = 1$ specifies a system of statistical isotropy usually associated with spherical inclusions. Intermediate values of $0 < a < \infty$ are of particular interest for semicrystalline polymers, particulate reinforced, and 'short' fibre reinforced systems. Changes of a in this region hold the potential for dramatic improvements in material performance through changes in the microstructure.[26]

The role of C^0_{ijkl} in eqn. (169) has been discussed previously. The following analysis, however, reveals additional physical insights. The model equation, (eqn. (169)), is a direct consequence of the stress polarisation field given by eqn. (164) (a piecewise constant function). In particular, the stress polarisation field in a region of phase i is a constant tensor given by:

$$p^i_{mn} = R^i_{mnkl}(\varepsilon^0_{kl} + E^0_{klop}p'^i_{op}) = R^i_{mnkl}(\varepsilon^0_{kl} - E^0_{klop}\langle p_{op}\rangle + E^0_{klop}p^i_{op}) \qquad (170)$$

In order to gain additional insights, it will be instructive to consider an ellipsoidal inclusion of phase i embedded in an infinite medium of elasticity C^0_{ijkl} with the strain tensor specified as a constant ε^∞_{ij} at infinity. The ellipsoidal inclusion is assumed to be of a shape consistent with the symmetry descriptors appearing in E^0_{klop}. This boundary value problem has been discussed by Eshelby;[13, 14] his solution shows that the inclusion is uniformly strained. (This result resembles the solutions of some well known electrostatic problems,[27] e.g. a dielectric inclusion placed in an initially uniform electric field.) In terms of the polarisation field, the solution within the inclusion can be cast into the following form (see Appendix 7):

$$p^i_{mn} = R^i_{mnkl}(\varepsilon^\infty_{kl} + E^0_{klop}p^i_{op}) \qquad (171)$$

and the uniform strain is given by:

$$\varepsilon_{kl}^i = \varepsilon_{kl}^\alpha + E_{klop}^0 p_{op}^i \tag{172}$$

For an ensemble of various phases subject to the same situation, the average strain of the ensemble is given by

$$\varepsilon_{kl}^0 = \langle \varepsilon_{kl} \rangle = \varepsilon_{kl}^\alpha + E_{klop}^0 \langle p_{op} \rangle \tag{173}$$

Substituting this relation into eqn. (171) yields:

$$p_{mn}^i = R_{mnkl}^i(\varepsilon_{kl}^0 - E_{klop}^0 \langle p_{op} \rangle + E_{klop}^0 p_{op}^i)$$

which is identical to eqn. (170), the stress field that leads to the model equation, (eqn. (169)).

This analysis suggests that the model equation, (eqn. (169)) corresponds to the following situations:

(i)   For a system of statistical ellipsoidal symmetry, every material element sees around itself (in an averaged sense) an elastic field of the same spacial symmetry.

(ii) Every individual material element experiences a stress field as if it were surrounded by a homogeneous matrix of elasticity C_{ijkl}^0.

(iii) Equation (169) results from the ensemble average of all elements consistent with the volume fractions and orientation distributions of various phases.

The above analysis provides a view of the physical role of the reference elasticity for application of eqn. (169) as a model constitutive relationship. In an actual heterogeneous system, every material element is surrounded by a different and non-uniform elastic field. If the strain energy of this system can be predicted by the model equation, (eqn. (169)), this system may be effectively replaced by a hypothetical system in which every material element is surrounded by an uniform surrogate matrix characterised by the reference elasticity. Under this view of the model, the reference elasticity is associated with the average effect that the surrounding matrix exercises on the material element in a heterogeneous system.

C.   Criterion for Selecting the Reference Elasticity for Model Constitutive Relationships

Clearly, a central problem for the application of eqn. (169) as a constitutive model is the determination of the reference elasticity from known information. For this purpose, the concepts that underlie the well-known self-consistent method of micromechanics can also be used here to assist in establishing criteria for selecting C^0_{ijkl}. Accordingly, it is worthwhile to summarise briefly some important treatments employing this scheme.

The self-consistent method was originally proposed by Hershey[28] and Kröner[29] for polycrystalline aggregates of metals. These workers approximated the real system by considering a single crystal inclusion embedded in an infinite matrix of the same property as the effective elasticity of the polycrystal of random orientation. To ensure self-consistency, the orientation average of the inclusion stress was set equal to that of the polycrystal subject to the same overall strain. This resulted in an implicit equation for the effective elasticity of the polycrystal.

The same spirit was adopted later by Hill[30] in developing his self-consistent model for two-phase composite materials. He considered two-phase systems having the character of a continuous matrix and spherical (or aligned ellipsoidal) inclusions with (if anisotropic) their correspondent material (or crystallographic) axes aligned. Similar approaches were taken by Hill and other workers[6, 30-32] to consider inclusions of other shapes and alignments. All these treatments have a common feature, viz., the system considered is required to be of the structure of inclusions embedded in a continuous matrix. Also because of mathematical difficulties, inclusions are restricted to having an uniform shape of simple geometry. These limitations on the structure preclude these treatments from being applied to more complicated heterogeneous systems, e.g., composites of a continuous matrix and arbitrarily shaped inclusions, systems which do not have a continuous phase and partially crystalline polymers which have crystalline regions of varying shapes or super structures such as 'shish-kebabs'.[33, 34] Indeed, no theoretically justified model currently exists for these complicated yet commonly encountered heterogeneous systems. Fortunately, the model equation, (eqn. (169)), is capable of serving as a model for such systems.

It should be emphasised that eqn. (169) was derived theoretically through the incorporation of certain statistical symmetry descriptors. Individual phase region geometries were completely disregarded. Consequently, eqn. (169) should be applicable to any heterogeneous system as long as it possesses statistical homogeneity and certain symmetries. It should be clearly understood that in the previous discussion, the inclusion problem was used only as a reference to help illustrate the physical meaning of eqn. (169) and should not be regarded as central to the development of this model equation.

The notion of self-consistency can easily be adopted to the present general model in order to furnish criteria for the selection of the reference elasticity. As illustrated previously, the reference elasticity may be interpreted as the average property of the matrix that the material elements experience. Accordingly, applying the self-consistent notion to this model suggests that the macroscopic effective moduli should be chosen as the reference elasticity. In view of eqns. (166) and (169), this implies that $\langle p_{ij}^t \rangle$ should be set equal to zero in order to ensure self-consistency. Specifically, averaging of eqn. (164) and setting $\langle p_{ij}^t \rangle$ equal to zero yields the following implicit equation for C_{ijkl}^0:

$$\langle M_{ijkl} \rangle = \langle (H_{ijkl} - E_{ijkl}^0)^{-1} \rangle = 0 \tag{174}$$

This result yields a new self-consistent model for general heterogeneous systems of statistical ellipsoidal symmetry: the effective elasticity is given by the C_{ijkl}^0 that satisfies eqn. (174).

For the purpose of illustration, consider a statistically isotropic system composed of two isotropic phases. Since the system is expected to behave isotropically on the whole, the self-consistent choice of C_{ijkl}^0 should be of isotropic symmetry. In the case of statistical isotropy and isotropic reference elasticity, the tensor quantity, E_{ijkl}^0, is given by eqn. (129), viz.,

$$E_{ijkl}^0 = (k^E - \tfrac{2}{3}\mu^E)\delta_{ij}\delta_{kl} + 2\mu^E I_{ijkl} \tag{175}$$

where

$$k^E = -\frac{1}{3(3k^0 + 4\mu^0)} \tag{176}$$

$$\mu^E = -\frac{3(k^0 + 2\mu^0)}{10\mu^0(3k^0 + 4\mu^0)} \tag{177}$$

with μ^0 the shear modulus and k^0 the bulk modulus of the reference elasticity. The volume fractions of the two phases are denoted by v_1 and v_2 and the respective phase properties are given by:

$$C_{ijkl}^n = (k^n - \tfrac{2}{3}\mu^n)\delta_{ij}\delta_{kl} + 2\mu^n I_{ijkl} \tag{178}$$

with $n = 1$ or 2. The correspondent H^n is easily found to be:

$$H_{ijkl}^n = (k^{H^n} - \tfrac{2}{3}\mu^{H^n})\delta_{ij}\delta_{kl} + 2\mu^{H^n} I_{ijkl} \tag{179}$$

where

$$k^{H^n} = \frac{1}{9(k^n - k^0)} \tag{180}$$

$$\mu^{H^n} = \frac{1}{4(\mu^n - \mu^0)} \tag{181}$$

Similarly, for each phase:

$$M^n_{ijkl} = \left[\frac{1}{9(k^{H''} - k^E)} - \frac{2}{3} \frac{1}{4(\mu^{H''} - \mu^E)} \right] \delta_{ij}\delta_{kl} + 2\frac{1}{4(\mu^{H''} - \mu^E)} I_{ijkl} \qquad (182)$$

In order to satisfy eqn. (174) for a self-consistent choice of C^0_{ijkl}, the volume average of M^n_{ijkl} must be set equal to zero. This implies that:

$$v_1(k^{H^2} - k^E) + v_2(k^{H^1} - k^E) = 0 \qquad (183)$$

and

$$v_1(\mu^{H^2} - \mu^E) + v_2(\mu^{H^1} - \mu^E) = 0 \qquad (184)$$

have to be satisfied simultaneously. In view of eqns. (180), (181), (176), and (177), these two eqns. can be written as

$$\frac{v_1}{k^2 - k^0} + \frac{v_2}{k^1 - k^0} = -\frac{3}{3k^0 + 4\mu^0} \qquad (185)$$

$$\frac{v_1}{\mu^2 - \mu^0} + \frac{v_2}{\mu^1 - \mu^0} = -\frac{6(k^0 + 2\mu^0)}{5\mu^0(3k^0 + 4\mu^0)} \qquad (186)$$

This result is exactly equivalent to Hill's results for composites composed of a continuous matrix with spherical inclusions.[30]

It is interesting to notice that both phases enter eqns. (185) and (186) on the same footing. This symmetry in the roles of the two phases is not *a priori* to the matrix-inclusion systems; such a symmetry implies that the same moduli will be predicted if the roles of the phases as continuous matrix and inclusions are interchanged. In the current case, however, such a symmetry is expected because, except for the statistical isotropy, nothing else concerning the geometrical structure was presumed. In view of this symmetry in the final equations, it is clear that the special reduction of the self-consistent model, viz., eqns. (185) and (186), and consequently Hill's model, should be restricted to systems in which the various components play roles of similar or equivalent significance.

In addition to the self-consistent method, there are other intuitive means to select the reference elasticity. The justification of a certain choice of the reference elasticity depends on the individual characteristics of the system of interest. When such a justification is difficult, experiments can be used for this determination. Indeed, both the reference elasticity and the statistical symmetry descriptors in the model equation, (eqn. (169)), can be treated as adjustable parameters subject to experimental correlations.

It should be noted that eqn. (169) is based on the assumption of statistical ellipsoidal symmetry. Accordingly, in order to use this equation to establish rigorous bounds, the system of interest must be known to possess such a symmetry. Otherwise, the results can only be regarded as 'model bounds', as discussed at the beginning of this section. However, if eqn. (169) is restricted in its use to a specific model, the statistical ellipsoidal symmetry does not necessarily have to be precisely observed in order to establish the status of its application as a model.

It is anticipated that most of the heterogeneous systems tend to have statistics of smoothly varying directional dependence of relatively high spacial symmetry. It is also anticipated that in most cases, this symmetry might well be ellipsoidal or can be reasonably approximated by ellipsoidal symmetry. Accordingly, the general model equation, (eqn. (169)), and the self-consistent form of eqn. (174), should suffice to serve as models for most of the important engineering applications.

VI. CONCLUSIONS

This treatment was aimed at two major objectives. The first objective involved the development of combining rules to predict overall effective properties of heterogeneous materials. Prior to this study, the Reuss and Voigt models were the only models which were applicable to heterogeneous materials, in general. However, these primitive averaging and combining rules are so inaccurate in many cases that they should not be used for any other purpose than obtaining rough estimates. Indeed, the Reuss and Voigt models correspond to the extreme bounds on composite properties; they are incapable of characterising material behaviours inbetween these extremes. Previous improvements over these combining rules, either in the form of rigorous bounds or micromechanical models, have been restricted to a few simple types of heterogeneous systems.

The second objective was to identify appropriate structural descriptors which dominate a wide spectrum of structure–property relationships. It was anticipated that these descriptors can serve as material design variables, which may be subject to control by the proper selection of processing conditions. In order to accomplish this objective, a combining rule general enough to deal with a broad spectrum of microstructures was required. Again, of the existing theories and constitutive models, only the Reuss and Voigt models were amenable to general applications. As emphasised previously, these models can only provide extreme bounds on the effective properties in terms of such primitive structural descriptors as volume fractions and orientation distributions. The fact that these bounds are impractically far apart suggests that additional structural descriptors play a major role in establishing the performance of heterogeneous materials. Past attempts to develop improvements over these primitive rules, by incorporating additional information, have only been formulated for a few special cases, which correspond to fragments of the observed spectrum of microstructures available to heterogeneous systems.

It was recognised that under the current directions of experimental studies and data analyses, only a limited amount of structural information concerning heterogeneous systems is accessible. Accordingly, it was natural to resort to variational principles in order to accommodate these partial descriptions of the microstructure. However, none of the existing forms of the classical variational principles possess the features required for the general treatment of these complicated systems. Consequently, new and improved forms of variational theorems were developed. The final optimal form is summarised in Theorem 4.

The application of the new variational theorems led to a general bounding theory capable of providing successively closer bounds by the systematic incorporation of well defined structural descriptors. In recognition of the limited amount and restricted forms of accessible information, the theory was cast in a form which isolates the roles of the dominant structural descriptors: volume fractions, orientation distributions, and statistical symmetry. The statistical symmetry parameters, which measure the morphological preferences along various directions, emerged as a significant structural descriptor of particular importance for semi-crystalline polymers, particulate reinforced, and 'short' fibre-reinforced systems. All other structural features are either prohibitively difficult to measure or their roles in determining the effective properties are not as yet clearly identified. The contributions of the remaining structural features are accordingly merged into the specification of a 'reference elasticity'.

The resulting constitutive relationship (eqn. (169)) provides a dual service. When all the other structural information is unknown, it provides rigorous bounds on the effective moduli by identifying the proper choice for the reference elasticity. The distance between the bounds can be viewed as a measure of the maximum influence that the unknown part of the structural information can possibly make. This new relationship can also serve either as a theoretical or a semiempirical model depending on the means used to determine the reference elasticity and, consequently, characterises all the structural features that have not been specifically taken into account.

This relationship provides improved bounds, as well as the basis for a more general constitutive model, for a much broader range of microstructures than accounted for in previous treatments. As was anticipated, volume fractions and orientation distributions play an important role in determining effective properties. The identification of the important role of statistical symmetry (as indicated by the averaged morphology of the reinforcing phase) provides a new material design variable to improve, control, and optimise the performance of heterogeneous materials.

This treatment has been applied to the analysis and prediction of the properties for a wide range of heterogeneous materials systems, viz., polycrystalline metals, fibre-reinforced composites, and partially crystalline polymers.[26] It is reasonable to believe that an even wider demonstration of applicability will be achieved as appropriate mechanical and structural data become available.

The fact that both structural and mechanical descriptors are essential to the rational analysis and interpretation of the behaviour of heterogeneous materials emphasises the necessity for combining structural experiments with the traditional mechanical property characterisation techniques. The significant and coupled roles of (i) volume fraction concentration, (ii) characteristics of the orientation distribution, and (iii) characteristics of the statistical symmetry along with the tensorial nature of the mechanical

TABLE 1

LINEAR TRANSPORT PROCESSES OF EQUIVALENT MATHEMATICAL STRUCTURE FOR THE
DETERMINATION OF EFFECTIVE TRANSPORT PROPERTIES
Local governing equations: $\mathbf{F} = \mathbf{K} \cdot \mathbf{G}$
(in steady state) $\nabla \cdot \mathbf{F} = 0$

Properties	F	G
Thermal conductivity	Heat flux	Temperature gradient
Electrical conductivity	Electrical current	Electrical field
Diffusivity	Molecular flux	Chemical potential gradient[a]
Dielectric constant	Electrical displacement	Electrical field
Magnetic permeability	Magnetic induction	Magnetic field
Linear elasticity	Stress field	Strain field

[a] The equilibrium condition presumed to exist at the phase boundary requires that the chemical potential, instead of the concentration, be continuous at the interface. It is thus more convenient to treat the chemical potential, other than the concentration, as the field variable.

properties, point out the futility of those experimental approaches which hope to describe and correlate the complex and subtle behaviour of heterogeneous materials in terms of empirical models based on one or two ill defined adjustable parameters.

It is of considerable significance that the treatments developed in this work can be easily applied to a wide range of properties for heterogeneous materials. In particular, a broad group of transport properties, as illustrated in Table 1, is governed by the same set of differential equations. The only major mathematical difference in the description of these properties is the tensor rank of the equations and the quantities involved in the various processes. These differences, of course, do not affect the equivalence of the mathematical structure of the governing equations as well as their solutions. Consequently, the treatments evolved in this work provide a basis for developing constitutive relationships for all of the properties listed in Table 1. Indeed, the linear elastic problem treated in this work contains quantities of higher tensor ranks than the correspondent quantities involved in any of the other processes listed in Table 1. Considerable simplifications in the formulation will be obtained for the other processes.

REFERENCES

1. HASHIN, Z. and SHTRIKMAN, S. *J. Mech. Phys. Solids*, **10** (1962) p. 335.
2. HASHIN, Z. and SHTRIKMAN, S. *J. Mech. Phys. Solids*, **10** (1962) p. 343.
3. HASHIN, Z. and SHTRIKMAN, S. *J. Mech. Phys. Solids*, **11** (1963) p. 127.
4. WALPOLE, L. J. *J. Mech. Phys. Solids*, **14** (1966) p. 151.
5. WALPOLE, L. J. *J. Mech. Phys. Solids*, **14** (1966) p. 289.
6. WALPOLE, L. J. *J. Mech. Phys. Solids*, **17** (1969) p. 235.
7. SOKOLNIKOFF, I. S. *Mathematical Theory of Elasticity*, McGraw-Hill, 1946.
8. LOVE, A. E. H. *A treatise on the mathematical theory of elasticity*, Cambridge Univ. Press, 1927.
9. PEARSON, C. E. *Theoretical elasticity*, Harvard Univ. Press, 1959.
10. COURANT, R. and HILBERT, D. Methods of mathematical physics, Vol. 1, Interscience 1953.
11. VOIGT, W. *Annln. Phys.*, **38** (1889) p. 573.
12. REUSS, A. *Z. Angew. Math. Mech.*, **9** (1929) p. 49.
13. ESHELBY, J. D. *Proc. Roy. Soc.*, **A241** (1957) p. 376.
14. ESHELBY, J. D. *Progress in soil mechanics* (Ed. I. N. Sneddon and R. Hill), Vol 2, Chap. III, North-Holland, 1961.
15. HASHIN, Z. *Int. J. Engng. Science*, **5,** (1967) p. 23.
16. SEEGER, A. *Handbuch d. Physik*, **7**(1) (1955) p. 383.
17. GREENBERG, M. D. *Application of Green's functions in science and engineering*, Prentice-Hall, 1971.

18. Lifshits, I. M. and Rozentsweig, L. N. *Z. Eksperim. i Theor. Phys.*, **17** (1947) p. 783.
19. Kröner, E. *Z. Physik*. **136** (1953) p. 402.
20. Chou, T. W. and Pan, Y. C. *J. Appl. Phys.*, **44** (1973) p. 63.
21. Leibfried, G. *Z. Physik*, **135** (1953) p. 23.
22. Bochner, S. and Chandrasekharan, K. *Fourier transforms*, Princeton Univ. Press, 1949.
23. Sneddon, I. N. *Fourier transforms*, McGraw-Hill, New York, 1951.
24. Ashton, J. E., Halpin, J. C. and Petit, P. H. *Primer on composite materials: Analysis*, Technomic, Stamford, Conn., 1969.
25. Kröner, E. *J. Mech. Phys. Solids*, **15** (1967) p. 319.
26. Wu, C.-T. D. Ph. D. Thesis, University of Delaware, Newark, Delaware, 1976.
27. Jackson, J. D. *Classical electrodynamics*, John Wiley & Sons, (1962).
28. Hershey, A. V. *J. Appl. Mech.*, **21** (1954) p. 236.
29. Kröner, E. *Z. Physik*, **151** (1958) p. 504.
30. Hill, R. *J. Mech. Phys. Solids*, **13** (1965) p. 213.
31. Hill, R. *J. Mech. Phys. Solids*, **13** (1965) p. 189.
32. Wu, T. T. *Int. J. Solids Struct.*, **2** (1966) p. 1.
33. Pennings, A. J., van der Mark, J. M. A. A. and Kiel, A. M. *Kolloid-Z.*, **237** (1970) p. 336.
34. Schultz, J. M. *Polymer materials science*, Prentice-Hall, 1974.
35. Hashin, Z. *J. Mech. Phys. Solids*, **13** (1965) p. 119.
36. Goldstein, H. Classical mechanics, Addison-Wesley, 1950.
37. Roe, R. J. and Krigbaum, W. R. *J. Chem. Phys.*, **40** (1964) p. 2608.
38. Hermans, P. H. Contributions to the physics of cellulose fibres, Elsevier, Amsterdam, 1946.

APPENDIX 1: APPLICATION OF VARIATIONAL PRINCIPLES TO HETEROGENEOUS SYSTEMS

The principle of minimum potential energy given by eqns. (1) to (4) can be applied to heterogeneous systems by modifying the subsidiary condition eqn. (3) as:

$$\varepsilon_{ij} = \tfrac{1}{2}(u_{i,j} + u_{j,i}) \quad \text{within phase regions; and } u_i\text{s are continuous at phase interfaces.} \quad \text{(A1-1)}$$

With this modification, the variational principle can be proven to generate the correct stationary condition required for a system at equilibrium.

First, with the help of eqn. (2), the objective (eqn. (1)) is rewritten as:

$$U_0 = \tfrac{1}{2} \int \varepsilon_{ij} C_{ijkl} \varepsilon_{kl} \, dV \quad \text{(A1-2)}$$

Taking the variation of eqn. (A1-2) yields:

$$\partial U_0 = \int \varepsilon_{ij} C_{ijkl} \, \partial \varepsilon_{kl} \, dV$$

or

$$\partial U_0 = \int \sigma_{ij}(\partial u_i)_{,j}\, \mathrm{d}V \qquad (A1\text{-}3)$$

The stationary condition can be found by setting ∂U_0 equal to zero for all small variations of the displacement field such that $u_i + \partial u_i$ satisfies the constraints imposed on u_i. Consequently, the variation field ∂u_i should satisfy the constraints that: (i) ∂u_is be continuous and (ii) $\partial u_i = 0$ on S.

In order to apply the divergence theorem to eliminate the differential operator on ∂u_i, it is necessary to replace the integration in eqn. (A1-3) by a summation of integrals over individual phase regions, viz.,

$$\partial U_0 = \sum_m \int \sigma_{ij}^m(\partial u_i^m)_{,j}\, \mathrm{d}V_m \qquad (A1\text{-}4)$$

where V_m denotes the volume of phase region m bounded by surface S_m and the summation is taken over all the phase regions bounded by surface S.

In every phase region, the elasticity field is assumed to be homogeneous and hence the divergence theorem is applicable. Accordingly, eqn. (A1-4) can be replaced by:

$$\partial U_0 = \sum_m \left(\int \sigma_{ij}^m\, \partial u_i^m n_j\, \mathrm{d}S_m - \int \sigma_{ij,j}^m\, \partial u_i^m\, \mathrm{d}V_m \right) \qquad (A1\text{-}5)$$

The stationary condition corresponds to the solution of the above equation with $\partial U_0 = 0$ for all continuous ∂u_i field such that $\partial u_i = 0$ on S. Consequently, it is evident that in order to generate the stationary solution, the following conditions must be met:

$$\sigma_{ij,j}^m = 0 \qquad \text{in } V_m \qquad (A1\text{-}6)$$

and $\sigma_{ij}^m n_j = \sigma_{ij}^n n_j$ or the forces balanced at the interface between phases m and n. This completes the proof.

Similarly, all the other variational principles developed in this work can be proven to be directly applicable to heterogeneous systems as long as the displacement field is required to be continuous.

APPENDIX 2: PROOF OF THEOREM 1

The stationary condition (eqn. (47)) can be proved by taking the variation of the objective (eqn. (54)). This gives:

$$\partial U_h = -\int (H_{ijkl}p_{kl}\, \partial p_{ij} - \varepsilon_{ij}^0\, \partial p_{ij} - \tfrac{1}{2}\varepsilon_{ij}'\, \partial p_{ij} - \tfrac{1}{2}p_{ij}\, \partial \varepsilon_{ij}')\, \mathrm{d}V \qquad (A2\text{-}1)$$

in which ∂p_{ij} and $\partial \varepsilon'_{ij}$ are required to be admissible to the set of constraints, viz., eqns. (46), (48) and (49). Note that the last term in eqn. (A2-1) can be written as

$$\int p_{ij}\, \partial \varepsilon'_{ij}\, \mathrm{d}V = \int (t_{ij} - C^0_{ijkl}\varepsilon'_{kl})\, \partial \varepsilon'_{ij}\, \mathrm{d}V \qquad (A2-2)$$

where t_{ij} is defined by eqn. (51). Clearly, both t_{ij} and ∂t_{ij} which is related to ∂p_{ij} and $\partial \varepsilon'_{ij}$ through eqn. (51), should satisfy the condition (52). In view of eqn. (53) and the constraints, (eqns. (48) and (49)) which the $\partial \varepsilon'_{ij}$ field is required to satisfy, it is evident that eqn. (A2-2) can be written as:

$$\int p_{ij}\, \partial \varepsilon'_{ij}\, \mathrm{d}V = - \int C^0_{ijkl}\varepsilon'_{kl}\, \partial \varepsilon'_{ij}\, \mathrm{d}V \qquad (A2-3)$$

With the help of eqn. (51), eqn. (A2-3) can be rewritten as:

$$\int p_{ij}\, \partial \varepsilon'_{ij}\, \mathrm{d}V = - \int \varepsilon'_{ij}(\partial t_{ij} - \partial p_{ij})\, \mathrm{d}V \qquad (A2-4)$$

In consequence of eqn. (53), the term $\int \varepsilon'_{ij}\, \partial t_{ij}\, \mathrm{d}V$ vanishes and eqn. (A2-4) reduces to

$$\int p_{ij}\, \partial \varepsilon'_{ij}\, \mathrm{d}V = \int \varepsilon'_{ij}\, \partial p_{ij}\, \mathrm{d}V \qquad (A2-5)$$

Substituting eqn. (A2-5) into (A2-1) leads to:

$$\partial U_h = - \int (H_{ijkl}p_{kl} - \varepsilon^0_{ij} - \varepsilon'_{ij})\, \partial p_{ij}\, \mathrm{d}V \qquad (A2-6)$$

By demanding $\partial U_h = 0$ for all ∂p_{ij}, the stationary condition is obtained as:

$$H_{ijkl}p_{kl} = \varepsilon^0_{ij} + \varepsilon'_{ij} \qquad (A2-7)$$

which is equivalent to eqn. (47).

The statements (55) and (56) can be proved by taking the second variation of U_h. This gives

$$\partial^2 U_h = \int (-H_{ijkl}\, \partial p_{ij}\, \partial p_{kl} + \partial \varepsilon'_{ij}\, \partial P_{ij})\, \mathrm{d}V \qquad (A2-8)$$

In view of eqn. (A2-3), the following relation can be obtained immediately:

$$\int \partial p_{ij}\, \partial \varepsilon'_{ij}\, \mathrm{d}V = - \int \partial \varepsilon'_{ij} C^0_{ijkl}\, \partial \varepsilon'_{kl}\, \mathrm{d}V \qquad (A2-9)$$

Substituting eqn. (A2-9) into (A2-8) yields

$$\partial^2 U_h = - \int (\partial p_{ij} H_{ijkl}\, \partial p_{kl} + \partial \varepsilon'_{ij} C^0_{ijkl}\, \partial \varepsilon'_{kl})\, \mathrm{d}V \qquad (A2-10)$$

Note that C^0_{ijkl} is always positive definite and for positive definite R_{ijkl}, the correspondent H_{ijkl} is also positive definite. Consequently, for all ∂p_{ij} and $\partial \varepsilon'_{ij}$,

$$\partial^2 U_h < 0 \qquad (A2-11)$$

if R_{ijkl} is kept positive definite. Accordingly, the resulting stationary solution corresponds to a maximum, and hence, statement (55) is proved. Alternately, the last term in eqn. (A2-8) can be written as

$$\int \partial \varepsilon'_{ij} \partial p_{ij} \, dV = \int (\partial t_{ij} S^0_{ijkl} \partial t_{kl} - \partial p_{ij} S^0_{ijkl} \partial p_{kl}) \, dV \qquad \text{(A2-12)}$$

with the help of the subsidiary conditions and eqns. (51) to (53). Substituting eqn. (A2-12) into (A2-8) yields

$$\partial^2 U_h = \int [-\partial p_{ij}(H_{ijkl} + S^0_{ijkl}) \partial p_{kl} + \partial t_{ij} S^0_{ijkl} \partial t_{kl}] \, dV \qquad \text{(A2-13)}$$

In view of the following identities:

$$(H_{ijkl} + S^0_{ijkl})R_{klmn} = I_{ijmn} + S^0_{ijkl}R_{klmn}$$

$$= S^0_{ijkl}C^0_{klmn} + S^0_{ijkl}R_{klmn}$$

$$= S^0_{ijkl}C_{klmn} \qquad \text{(A2-14)}$$

it is evident that the negative definiteness of R_{ijkl} implies the negative definiteness of $H_{ijkl} + S^0_{ijkl}$. Accordingly, if R_{ijkl} is kept negative definite,

$$\partial^2 U_h > 0 \qquad \text{(A2-15)}$$

for all ∂p_{ij} and ∂t_{ij}. This implies that the resulting stationary solution corresponds to a minimum for negative definite R_{ijkl}. Hence, statement (56) is proved.

This completes the proof of Theorem 1. Theorems 2, 3, and 4 can be similarly proved.

APPENDIX 3: COMMUTATION BETWEEN DIFFERENTIAL AND AVERAGING OPERATORS

The two-point correlation function $\langle p_{ij,j}(\mathbf{r}')p_{kl,l}(\mathbf{r}' - \mathbf{r})\rangle$ can be expressed in terms of the correlation of p_{ij} field by properly commuting the differential and averaging operators.

Note that the term $p_{kl,l}(\mathbf{r}' - \mathbf{r})$ can be written as $-(\partial/\partial x_l)p_{kl}(\mathbf{r}' - \mathbf{r})$ because:

$$-\frac{\partial}{\partial x_l} p_{kl}(\mathbf{r}' - \mathbf{r}) = -p_{kl,l}(\mathbf{r}' - \mathbf{r})\frac{\partial}{\partial x_l}(x'_l - x_l)$$

$$= p_{kl,l}(\mathbf{r}' - \mathbf{r}) \qquad \text{(A3-1)}$$

In consequence of eqn. (A3-1), it immediately follows that:

$$\langle p_{ij,j}(\mathbf{r}')p_{kl,l}(\mathbf{r}'-\mathbf{r})\rangle = -\langle p_{ij,j}(\mathbf{r}')\frac{\partial}{\partial x_l}p_{kl}(\mathbf{r}'-\mathbf{r})\rangle$$

$$= -\frac{\partial}{\partial x_l}\langle p_{ij,j}(\mathbf{r}')p_{kl}(\mathbf{r}'-\mathbf{r})\rangle \qquad (A3\text{-}2)$$

since the averaging operation is taken over the $\mathbf{r}'$ space and is thence commutative with the differentiation $(\partial/\partial x_l)$ in the $\mathbf{r}$ space.

By using the same arguments that lead to eqns. (A3-1) and (A3-2), it can be readily shown that:

$$\langle p_{ij,j}(\mathbf{r}')p_{kl}(\mathbf{r}'-\mathbf{r})\rangle = \langle p_{ij,j}(\mathbf{r}'+\mathbf{r})p_{kl}(\mathbf{r}')\rangle$$

$$= \langle \frac{\partial}{\partial x_j}p_{ij}(\mathbf{r}'+\mathbf{r})p_{kl}(\mathbf{r}')\rangle$$

$$= \frac{\partial}{\partial x_j}\langle p_{ij}(\mathbf{r}'+\mathbf{r})p_{kl}(\mathbf{r}')\rangle$$

$$= \frac{\partial}{\partial x_j}\langle p_{ij}(\mathbf{r}')p_{kl}(\mathbf{r}'-\mathbf{r})\rangle \qquad (A3\text{-}3)$$

Substituting eqn. (A3-3) into (A3-2) leads to the following relationship:

$$\langle p_{ij,j}(\mathbf{r}')p_{kl,l}(\mathbf{r}'-\mathbf{r})\rangle = -\frac{\partial^2}{\partial x_j \partial x_l}\langle p_{ij}(\mathbf{r}')p_{kl}(\mathbf{r}'-\mathbf{r})\rangle \qquad (A3\text{-}4)$$

APPENDIX 4: DETERMINATION OF $g_{ij}^0(\mathbf{r})$ AND $G_{ijkl}^0(\mathbf{r})$ FROM THE FOURIER TRANSFORMS

The Green's tensor function $g_{ij}^0(\mathbf{r})$ and the correlation weighting function $G_{ijkl}^0(\mathbf{r})$ for isotropic media can be constructed by taking the inverse transformation of their Fourier transforms, viz., eqns. (114) and (118). It is most convenient to start with the integral relation:

$$\frac{1}{\pi^2}\int\frac{\exp(i\mathbf{k}\cdot\mathbf{r})}{k^4}dV_k = -r \qquad (A4\text{-}1)$$

which can be obtained by direct integration over the 'k' space.

Successive differentiation of eqn. (A4-1) with respect to x_i and x_j leads to the following relation:

$$-\frac{1}{\pi^2}\int\frac{k_i k_j \exp(i\mathbf{k}\cdot\mathbf{r})}{k^4}dV_k = -r_{,ij} \qquad (A4\text{-}2)$$

and hence

$$\frac{1}{\pi^2} \int \frac{\exp(i\mathbf{k} \cdot \mathbf{r})}{k^2} dV_k = r_{,mm} \tag{A4-3}$$

Note that

$$r_{,ij} = \frac{1}{r}(\delta_{ij} - x_i x_j / r^2) \tag{A4-4}$$

and hence

$$r_{,mm} = 2/r \tag{A4-5}$$

Taking the inverse transformation of eqn. (114) and using the relations (A4-2) through (A4-5) immediately leads to eqn. (116).

Successively differentiating (A4-2) and (A4-3) with respect to x_k and x_l yields:

$$\frac{1}{\pi^2} \int \frac{k_i k_j k_k k_l \exp(i\mathbf{k} \cdot \mathbf{r})}{k^4} dV_k = -r_{,ijkl} \tag{A4-6}$$

and

$$\frac{1}{\pi^2} \int \frac{k_k k_l \exp(i\mathbf{k} \cdot \mathbf{r})}{k^2} dV_k = -r_{,mmkl} \tag{A4-7}$$

respectively. Note that:

$$r_{,ijkl} = \frac{1}{r^3}[3(\delta_{kl} x_i x_j + \delta_{jl} x_i x_k + \delta_{jk} x_i x_l + \delta_{il} x_j x_k + \delta_{ij} x_k x_l + \delta_{ik} x_l x_j)/r^2$$

$$- (\delta_{ij}\delta_{kl} + \delta_{il}\delta_{kj} + \delta_{jl}\delta_{ik}) - 15 x_i x_j x_k x_l / r^4] \tag{A4-8}$$

$$r_{,mmkl} = \frac{1}{r^3}(6 x_k x_l / r^2 - 2\delta_{kl}) \tag{A4-9}$$

Taking the inverse transformation of eqn. (118) and using the relations (A4-6) to (A4-9) immediately leads to eqn. (119).

APPENDIX 5: CONSTRUCTION OF E^0_{ijkl} FOR ELLIPSOIDAL SYMMETRY

Consider the special case of isotropic reference elasticity and ellipsoidal symmetry with $a_1 = a_2 = 1$ and $a_3 = a$. Thus, $\bar{G}^0_{ijkl}$, defined by eqn. (144) is given by:

$$\bar{G}^0_{ijkl} = \frac{\lambda^0 + \mu^0}{\mu^0(\lambda^0 + 2\mu^0)} \frac{\zeta_i \zeta_j \zeta_k \zeta_l}{\zeta^4} - \frac{1}{4\mu^0 \zeta^2}(\delta_{ik}\zeta_j\zeta_l + \delta_{jk}\zeta_i\zeta_l + \delta_{il}\zeta_j\zeta_k + \delta_{jl}\zeta_i\zeta_k)$$

$$\tag{A5-1}$$

where

$$\zeta_1 = \cos\varphi \sin\eta$$
$$\zeta_2 = \sin\varphi \sin\eta$$
$$\zeta_3 = \cos\eta/a \qquad \text{(A5-2)}$$

and

$$\zeta^2 = \zeta_i\zeta_i = \sin^2\eta + \cos^2\eta/a^2 \qquad \text{(A5-3)}$$

E^0_{ijkl} is related to $\bar{G}^0_{ijkl}$ through eqn. (147). It is easy to verify that E^0_{ijkl} for the current case is a transversely isotropic tensor, and hence, can be denoted by:

$$E^0 = (k^E_T, \mu^E_T, \lambda^E_A, C^E_A, \mu^E_A) \qquad \text{(A5-4)}$$

Direct integration of eqn. (147) yields:

$$k^E_T = \frac{\lambda^0 + \mu^0}{\mu^0(\lambda^0 + 2\mu^0)}\frac{h_3(a)}{4} - \frac{1}{\mu^0}\frac{h_1(a)}{4}$$

$$\mu^E_T = \frac{\lambda^0 + \mu^0}{\mu^0(\lambda^0 + 2\mu^0)}\frac{h_3(a)}{8} - \frac{1}{\mu^0}\frac{h_1(a)}{4}$$

$$\lambda^E_A = \frac{\lambda^0 + \mu^0}{\mu^0(\lambda^0 + 2\mu^0)}\frac{h_4(a)}{2}$$

$$C^E_A = \frac{\lambda^0 + \mu^0}{\mu^0(\lambda^0 + 2\mu^0)}h_5(a) - \frac{1}{\mu^0}h_2(a)$$

$$\mu^E_A = \frac{\lambda^0 + \mu^0}{\mu^0(\lambda^0 + 2\mu^0)}\frac{h_4(a)}{2} - \frac{1}{4\mu^0}\left(\frac{h_1(a)}{2} + h_2(a)\right) \qquad \text{(A5-5)}$$

where $h_i(a)$s are defined as:

$$h_1(a) = \int_0^1 \frac{1 - y^2}{1 - by^2}\,dy$$

$$h_2(a) = \int_0^1 \frac{y^2}{a^2(1 - by^2)}\,dy$$

$$h_3(a) = \int_0^1 \frac{(1 - 2y^2 + y^4)}{(1 - by^2)^2}\,dy$$

$$h_4(a) = \int_0^1 \frac{y^2 - y^4}{a^2(1 - by^2)^2}\,dy$$

$$h_5(a) = \int_0^1 \frac{y^4}{a^4(1 - by^2)^2}\,dy \qquad \text{(A5-6)}$$

where b is defined as:

$$b = (a^2 - 1)/a^2$$

These integrations can be carried out analytically. The results are:

(i) For $0 \le a < 1$,

$$h_1(a) = \frac{a^2}{1 - a^2}\left\{\left[\left(\frac{1 - a^2}{a^2}\right)^{1/2} + \left(\frac{a^2}{1 - a^2}\right)^{1/2}\right]\tan^{-1}\left(\frac{1 - a^2}{a^2}\right)^{1/2} - 1\right\}$$

$$h_2(a) = \frac{1}{1 - a^2}\left\{1 - \left(\frac{a^2}{1 - a^2}\right)^{1/2}\tan^{-1}\left(\frac{1 - a^2}{a^2}\right)^{1/2}\right\}$$

$$h_3(a) = \left(\frac{a^2}{1 - a^2}\right)^2\left\{\frac{1}{2a^2} + 1 + \left[\frac{1}{2}\left(\frac{1 - a^2}{a^2}\right)^{3/2} - \left(\frac{1 - a^2}{a^2}\right)^{1/2}\right.\right.$$
$$\left.\left. - \frac{3}{2}\left(\frac{a^2}{1 - a^2}\right)^{1/2}\right]\tan^{-1}\left(\frac{1 - a^2}{a^2}\right)^{1/2}\right\}$$

$$h_4(a) = \frac{a^2}{(1 - a^2)^2}\left\{-\frac{3}{2} + \left[\frac{1}{2}\left(\frac{1 - a^2}{a^2}\right)^{1/2} + \frac{3}{2}\left(\frac{a^2}{1 - a^2}\right)^{1/2}\right]\right.$$
$$\left. \times \tan^{-1}\left(\frac{1 - a^2}{a^2}\right)^{1/2}\right\}$$

$$h_5(a) = \frac{1}{(1 - a^2)^2}\left\{1 + \frac{a^2}{2} - \frac{3}{2}\left(\frac{a^2}{1 - a^2}\right)^{1/2}\tan^{-1}\left(\frac{1 - a^2}{a^2}\right)^{1/2}\right\} \qquad \text{(A5-7)}$$

(ii) For $a = 1$,

$$h_1(1) = 2/3 \qquad h_2(1) = 1/3 \qquad h_3(1) = 8/15$$
$$h_4(1) = 2/15 \qquad h_5(1) = 1/5 \qquad\qquad \text{(A5-8)}$$

(iii) For $a > 1$,

$$h_1(a) = \frac{a^2}{a^2 - 1}\left\{1 - \frac{1}{2}\left[\left(\frac{a^2}{a^2 - 1}\right)^{1/2} - \left(\frac{a^2 - 1}{a^2}\right)^{1/2}\right]\ln\frac{a + (a^2 - 1)^{1/2}}{a - (a^2 - 1)^{1/2}}\right\}$$

$$h_2(a) = \frac{1}{a^2 - 1}\left[-1 + \frac{1}{2}\left(\frac{a^2}{a^2 - 1}\right)^{1/2}\ln\frac{a + (a^2 - 1)^{1/2}}{a - (a^2 - 1)^{1/2}}\right]$$

$$h_3(a) = \left(\frac{a^2}{a^2 - 1}\right)^2\left\{1 + \frac{1}{2a^2} + \left[\frac{1}{4}\left(\frac{a^2 - 1}{a^2}\right)^{3/2} + \frac{1}{2}\left(\frac{a^2 - 1}{a^2}\right)^{1/2}\right.\right.$$
$$\left.\left. - \frac{3}{4}\left(\frac{a^2}{a^2 - 1}\right)^{1/2}\right]\ln\frac{a + (a^2 - 1)^{1/2}}{a - (a^2 - 1)^{1/2}}\right\}$$

$$h_4(a) = \frac{a^2}{(a^2-1)^2}\left\{-\frac{3}{2} + \left[-\frac{1}{4}\left(\frac{a^2-1}{a^2}\right)^{1/2} + \frac{3}{4}\left(\frac{a^2}{a^2-1}\right)^{1/2}\right]\right.$$
$$\left. \times \ln\frac{a+(a^2-1)^{1/2}}{a-(a^2-1)^{1/2}}\right\}$$

$$h_5(a) = \frac{1}{(a^2-1)^2}\left\{1 + \frac{a^2}{2} - \frac{3}{4}\left(\frac{a^2}{a^2-1}\right)^{1/2}\ln\frac{a+(a^2-1)^{1/2}}{a-(a^2-1)^{1/2}}\right\} \qquad \text{(A5-9)}$$

APPENDIX 6: TENSORIAL TRANSFORMATIONS AND ORIENTATION AVERAGING

The components of the elasticity tensor (as well as other related tensor quantities) of a material are usually specified with respect to a specific Cartesian coordinate system, whose axes (commonly referred to as material axes) are associated with some natural directions unique to the material. These material axes do not, in general, coincide with the body axes system, in which the external loads (or deformations) are specified. Hence it is necessary to apply coordinate transformations to bring various tensor quantities into the common coordinate system associated with the body axes.

The body axes will be denoted by x_1, x_2, and x_3, and the material axes by $\bar{x}_1, \bar{x}_2$, and $\bar{x}_3$. Both are assumed to be Cartesian. The relative orientation between these two coordinate systems can be characterised by the Eulerian angles (φ, θ, ψ) as defined in Fig. 1. The orientation relationships between these two set of co-ordinates is given by

$$\bar{x}_i = \omega_{ij}x_j \qquad \text{(A6-1)}$$

where the transformation matrix ω_{ij} can be expressed in terms of the Eulerian angles as[36]

$$\omega_{ij} = \begin{bmatrix} \cos\psi\cos\varphi - \cos\theta\sin\varphi\sin\psi & \cos\psi\sin\varphi + \cos\theta\cos\varphi\sin\psi & \sin\psi\sin\theta \\ -\sin\psi\cos\varphi - \cos\theta\sin\varphi\cos\psi & -\sin\psi\sin\varphi + \cos\theta\cos\varphi\cos\psi & \cos\psi\sin\theta \\ \sin\theta\sin\varphi & -\sin\theta\cos\varphi & \cos\theta \end{bmatrix}$$
$$\text{(A6-2)}$$

Note that ω_{ij} is an orthogonal matrix, whose inverse can be immediately written as

$$\omega_{ij}^{-1} = \omega_{ij}^T = \omega_{ji} \qquad \text{(A6-3)}$$

Consider a fourth rank Cartesian tensor, $\bar{A}_{ijkl}$, whose components are

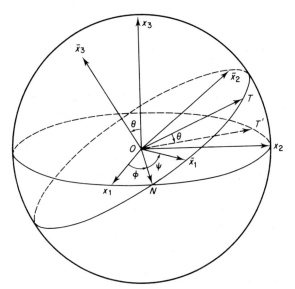

FIG. 1.   Eulerian angles.

specified with respect to the material axes $\bar{x}_i$s. Such a tensor follows the Cartesian transformation rule:

$$A_{mnop} = \frac{\partial x_m}{\partial \bar{x}_i} \frac{\partial x_n}{\partial \bar{x}_j} \frac{\partial x_o}{\partial \bar{x}_k} \frac{\partial x_p}{\partial \bar{x}_l} \bar{A}_{ijkl} \qquad (\text{A6-4})$$

where the A_{mnop}s are the components with respect to the body axes x_is. From eqns. (A6-1) and (A6-3), (A6-4) can be written as

$$A_{mnop} = \omega_{im}\omega_{jn}\omega_{ko}\omega_{lp}\bar{A}_{ijkl} \qquad (\text{A6-5})$$

In a heterogeneous system, the distinct phase regions can be distributed with respect to their orientations to the body axes. Accordingly, a weighted average of some tensor quantity (of fourth rank) over all possible orientations is required for the estimation of effective moduli. Clearly, in order to compute such an average, the orientation distribution (or certain characteristics of the distribution) must be specified.

Consider a collection of elements of property $\bar{A}_{mnop}$, whose orientations relative to the body axes are distributed according to some normalised orientation density function, $n(\varphi, \theta, \psi)$. From (A6-5), the average for this property is obtained as:

$$\langle A_{ijkl} \rangle_{\varphi, \theta, \psi} = \langle \omega_{mi}\omega_{nj}\omega_{ok}\omega_{pl} \rangle \bar{A}_{mnop} \qquad (\text{A6-6})$$

with

$$\langle \omega_{mi}\omega_{nj}\omega_{ok}\omega_{pl} \rangle = \int_0^{2\pi} \int_0^{\pi} \int_0^{2\pi} \omega_{mi}\omega_{nj}\omega_{ok}\omega_{pl} n(\varphi, \theta, \psi) \sin \theta \, d\varphi \, d\theta \, d\psi$$

(A6-7)

where $\langle A_{ijkl} \rangle_{\varphi,\theta,\psi}$ is the averaged property specified with respect to the body axes; the subscripts φ, θ, and ψ denote the average over a distribution in these angles.

Although more complicated orientation distributions can be anticipated, the special case of an axial orientation[37] is of particular significance. The axial orientation represents a situation in which a unique, say $\bar{x}_3$, axis of the material (e.g., the 'c' or chain axis in polymeric solids) is distributed according to a density function at an angle θ with respect to a body axis, say x_3 (e.g., the draw direction). The other two Eulerian angles are taken to be randomly distributed so that $n(\varphi, \theta, \psi)$ reduces to $n(\theta)$. Such a distribution

TABLE 2

CORRESPONDENCE BETWEEN THE VOIGT AND THE TENSOR INDICES

Voigt indices	m	1	2	3	4	5	6
Tensor indices	ij	11	22	33	23, 32	31, 13	12, 21

generates a transversely isotropic averaged tensor, which may be expressed in terms of the Voigt notation (Table 2) as

$$\langle A_{mn} \rangle_\theta = \begin{pmatrix} D_1 + D_2 & D_1 - D_2 & D_3 & & & \\ D_1 - D_2 & D_1 + D_2 & D_3 & & & \\ D_3 & D_3 & D_4 & & & \\ & & & D_5 & & \\ & & & & D_5 & \\ & & & & & D_2 \end{pmatrix}$$

(A6-8)

with all the other elements equal to zero. The terms D_is are given by

$$D_i = \sum_{j=1}^{5} (\alpha_{ij} + f\beta_{ij} + g\gamma_{ij}) \bar{I}_j$$

(A6-9)

with $i = 1, 2, \ldots, 5$.

The α_{ij}, β_{ij}, and γ_{ij} are constants which result from the random averaging over the angle φ and ψ. The 5×5 arrays for the $\boldsymbol{\alpha}$, $\boldsymbol{\beta}$, and γ are summarised

TABLE 3
ORIENTATION CONSTANTS FOR AXIAL DISTRIBUTIONS

$$\alpha = \frac{1}{15}\begin{pmatrix} 7 & 2 & 6 & 2 & 2 \\ 1 & 6 & -2 & 1 & 6 \\ 6 & -4 & 8 & 1 & -4 \\ 8 & 8 & 4 & 3 & 8 \\ 1 & 6 & -2 & 1 & 6 \end{pmatrix} \qquad \beta = \frac{1}{6}\begin{pmatrix} 2 & -2 & 0 & -2 & 4 \\ -1 & 3 & 2 & -1 & 0 \\ 0 & 4 & -2 & 2 & -8 \\ -8 & -8 & 8 & 0 & 16 \\ 2 & 0 & -4 & 2 & -6 \end{pmatrix}$$

$$\gamma = \frac{1}{10}\begin{pmatrix} 2 & 2 & -4 & 2 & -8 \\ 1 & 1 & -2 & 1 & -4 \\ -4 & -4 & 8 & -4 & 16 \\ 8 & 8 & -16 & 8 & -32 \\ -4 & -4 & 8 & -4 & 16 \end{pmatrix} \qquad \lambda = \frac{1}{8}\begin{pmatrix} 2 & 2 & 4 & 2 & 0 \\ 1 & 1 & -2 & 1 & 4 \\ 4 & -4 & 4 & 0 & 0 \\ 8 & 8 & 0 & 0 & 0 \\ 0 & 4 & 0 & 0 & 4 \end{pmatrix}$$

in Table 3. The orientation invariants, $\bar{I}_j$s, are defined in Table 4. The quantity 'f' is the familiar Hermans orientation 'function':[38]

$$f = \frac{1}{2}\left(3\int_0^{\pi} \cos^2\theta n(\theta)\sin\theta\,d\theta - 1\right) \qquad (A6\text{-}10)$$

The term g corresponds to a similar definition for the fourth moment of $\cos\theta$:

$$g = \frac{1}{2}\left(5\int_0^{\pi} \cos^4\theta n(\theta)\sin\theta\,d\theta - 1\right) \qquad (A6\text{-}11)$$

Both f and g have the property that for perfect alignment of the $\bar{x}_3$ axis with the body axis, x_3, $f = g = 1$; for a random orientation, $f = g = 0$. In the event that the unique material axis, $\bar{x}_3$, is perpendicular to the x_3 axis but randomly distributed in the transverse plane, $f = -\frac{1}{2}$ and $g = -\frac{1}{4}$. These relationships demonstrate that a complete knowledge of the orientation distribution is not required to describe the averaged tensor properties; characteristics of the distributions, such as f and g, yield sufficient structural information to determine averaged properties.

TABLE 4
INVARIANTS FOR AXIAL ORIENTATION DISTRIBUTIONS

$$\bar{I}_1 = (\bar{A}_{11} + \bar{A}_{22} + 2\bar{A}_{12})/4$$
$$\bar{I}_2 = (\bar{A}_{11} + \bar{A}_{22} - 2\bar{A}_{12} + 4\bar{A}_{66})/8$$
$$\bar{I}_3 = (\bar{A}_{13} + \bar{A}_{23})/2$$
$$\bar{I}_4 = \bar{A}_{33}$$
$$\bar{I}_5 = (\bar{A}_{44} + \bar{A}_{55})/2$$

In the event of perfect alignment ($f = g = 1$) the relationships reduce to $D_i = \bar{I}_i$ (since $\alpha_{ij} + \beta_{ij} + \gamma_{ij} = \delta_{ij}$). Accordingly, the invariants, $\bar{I}_i$, can be identified as the averaged tensor properties resulting from a random distribution around the coincident unique material and body axes. In the event that the unique material axis is perpendicular to the unique body axis ($f = -\frac{1}{2}$, $g = -\frac{1}{4}$), the relationships reduce to:

$$D_i = \sum_{j=1}^{5} (\alpha_{ij} - \tfrac{1}{2}\beta_{ij} - \tfrac{1}{4}\gamma_{ij})\bar{I}_j$$

or

$$D_i = \sum_{j=1}^{5} \lambda_{ij}\bar{I}_j \tag{A6-12}$$

The elements of λ are also given in Table 3.

In the event of random orientation ($f = g = 0$), eqn. (A6-9) reduces to:

$$D_i = \sum_{j=1}^{5} \alpha_{ij}\bar{I}_j \tag{A6-13}$$

It can be readily shown that this condition results in an isotropic averaged tensor $\langle A_{ijkl} \rangle$, which may be rearranged into the following familiar form:

$$\langle A_{ijkl} \rangle = \lambda^A \delta_{ij}\delta_{kl} + 2\mu^A I_{ijkl} \tag{A6-14}$$

with

$$\lambda^A = (B_1 + 4B_2 - 2B_3)/5 \tag{A6-15}$$

$$\mu^A = (B_1 - B_2 + 3B_3)/5 \tag{A6-16}$$

where B_1, B_2, and B_3 are related to $\bar{A}_{mnop}$ through

$$B_1 = (\bar{A}_{11} + \bar{A}_{22} + A_{33})/3 \tag{A6-17}$$

$$B_2 = (\bar{A}_{12} + \bar{A}_{23} + \bar{A}_{31})/3 \tag{A6-18}$$

$$B_3 = (\bar{A}_{44} + \bar{A}_{55} + \bar{A}_{66})/3 \tag{A6-19}$$

When $\bar{A}_{ijkl}$ is replaced by the elasticity tensor, these results reduce to the Voigt average for isotropic aggregates of crystals; when $\bar{A}_{ijkl}$ is replaced by the compliance tensor, the Reuss average is obtained.

APPENDIX 7: THE STRESS POLARISATION FIELD WITHIN AN ELLIPSOIDAL INCLUSION EMBEDDED IN AN INFINITE MEDIUM

The solution of the elastic problem of an ellipsoidal inclusion of elasticity C_{ijkl} embedded in an infinite medium of elasticity C^0_{ijkl} with the strain field specified as a constant ε^α_{ij} at infinity has been obtained by solving the set of differential equations (2) to (5), or (46) to (49), with the boundary conditions properly imposed.[13,14] This problem can also be solved through the integral equation (104) in a manner less straightforward yet relevant to the present work.

By taking the elasticity of the infinite medium as the reference, it immediately follows that the stress polarisation field outside the inclusion is identically zero. Thus, the prime concern will be to find the stress polarisation field within the inclusion, which is assumed to be bounded by the ellipsoid:

$$(x_1/a_1)^2 + (x_2/a_2)^2 + (x_3/a_3)^2 = 1 \qquad (A7-1)$$

To solve the integral equation (104), the method of successive approximation will be used. It is reasonable to select the initial approximation p^1_{ij} as follows:

$$p^1_{ij} = K_{ij} = \text{Constant} \qquad \text{within the inclusion}$$

and

$$p^1_{ij} = 0 \qquad \text{outside the inclusion} \qquad (A7-2)$$

The second approximation, if necessary, can be obtained by substituting (A7-2) into the right hand side of eqn. (104), viz.,

$$p^2_{ij} = R_{ijkl}(\varepsilon^\alpha_{kl} + \int G^0_{klmn}(\mathbf{r} - \mathbf{r}')p'^1_{mn}(\mathbf{r}')\,dV') \qquad (A7-3)$$

This procedure should be successively carried out until the approximating solutions converge. In view of eqn. (A7-3), it is necessary to evaluate the integral term, subsequently denoted by $Q_{ij}(\mathbf{r})$, on the right-hand side of the integral equation for the first approximation. In view of eqn. (96), Q_{ij} can be written as

$$Q_{ij}(\mathbf{r}) = (1/2\pi)^3 \iint \bar{G}^0_{ijkl}(\mathbf{k})p'_{kl}(\mathbf{r}')\exp(i\mathbf{k}\cdot\mathbf{r}' - i\mathbf{k}\cdot\mathbf{r})\,dV_k\,dV' \qquad (A7-4)$$

For convenience, the following coordinate transformations are introduced:

$$\rho_1 = x_1/a_1$$
$$\rho_2 = x_2/a_2$$
$$\rho_3 = x_3/a_3 \qquad (A7\text{-}5)$$

$$\rho'_1 = x'_1/a_1 = \cos\psi \sin\theta$$
$$\rho'_2 = x'_2/a_2 = \sin\psi \sin\theta$$
$$\rho'_3 = x'_3/a_3 = \cos\theta \qquad (A7\text{-}6)$$

$$s_1 = k_1 a_1 = \cos\varphi \sin\eta$$
$$s_2 = k_2 a_2 = \sin\varphi \sin\eta$$
$$s_3 = k_3 a_3 = \cos\eta \qquad (A7\text{-}7)$$

In terms of these new coordinates, eqn. (A7-2) can be restated as:

$$p^1_{ij} = K_{ij} \qquad \text{for } \rho < 1$$
$$= 0 \qquad \text{for } \rho > 1 \qquad (A7\text{-}8)$$

and $Q_{ij}(\mathbf{r})$ can be rewritten as

$$Q_{ij}(\boldsymbol{\rho}) = (1/2\pi)^3 \iint \bar{G}^0_{ijkl}(\varphi, \eta) p'_{kl}(\boldsymbol{\rho}') \exp(i\mathbf{s}\cdot\boldsymbol{\rho}' - i\mathbf{s}\cdot\boldsymbol{\rho})\, dV_s\, dV'_\rho \quad (A7\text{-}9)$$

where $\bar{G}^0_{ijkl}(\varphi, \eta)$ is given by eqn. (144), which is not a function of s as shown in Section III-D.

By substituting eqn. (A7-8) into (A7-9), Q_{ij} becomes:

$$Q_{ij}(\boldsymbol{\rho}) = \frac{K_{kl}}{8\pi^3} \int \bar{G}^0_{ijkl} \exp(-i\mathbf{s}\cdot\boldsymbol{\rho})\, dV_s \int_0^{2\pi}\int_0^\pi\int_0^1 \exp(i\mathbf{s}\cdot\boldsymbol{\rho}')\rho'^2 \sin\theta\, d\rho'\, d\theta\, d\psi$$

$$(A7\text{-}10)$$

It can be readily shown (through direct integration) that:

$$\int_0^{2\pi}\int_0^\pi\int_0^1 \exp(i\mathbf{s}\cdot\boldsymbol{\rho}')\rho'^2 \sin\theta\, d\rho'\, d\theta\, d\psi = 4\pi(\sin s/s^3 - \cos s/s^2)$$

$$(A7\text{-}11)$$

Substituting eqn. (A7-11) into (A7-10) yields:

$$Q_{ij}(\boldsymbol{\rho}) = \frac{K_{kl}}{8\pi^3} \int \bar{G}^0_{ijkl} \exp(-i\mathbf{s}\cdot\boldsymbol{\rho})4\pi(\sin s/s^3 - \cos s/s^2)\, dV_s$$

$$= \frac{K_{kl}}{2\pi^2} \int_0^{2\pi}\int_0^\pi \bar{G}_{ijkl} \sin\eta\, d\eta\, d\varphi \int_0^\infty \left(\frac{\sin s}{s} - \cos s\right)\exp(-i\mathbf{s}\cdot\boldsymbol{\rho})\, ds$$

$$(A7\text{-}12)$$

Note that the exponential term can be decomposed into a real part and an imaginary part. It is evident that the imaginary part will not contribute to the integral because of the symmetry of $\bar{G}^0_{ijkl}$. Accordingly, (A7-12) can be written as

$$Q_{ij}(\boldsymbol{\rho}) = \frac{K_{kl}}{2\pi^2} \int_0^{2\pi} \int_0^{\pi} \bar{G}^0_{ijkl}(\varphi, \eta) \sin \eta \, \mathrm{d}\eta \, \mathrm{d}\varphi$$

$$\times \int_0^{\infty} \left(\frac{\sin s}{s} - \cos s\right) \cos(s\rho \cos \beta) \, \mathrm{d}s \qquad \text{(A7-13)}$$

where β denotes the angle between the vectors $\mathbf{s}$ and $\boldsymbol{\rho}$. Note that within the inclusion $\rho < 1$ and $|\cos \beta|$ is bounded by 1, and consequently, $|\rho \cos \beta| < 1$. It can be readily shown that for $|\rho \cos \beta| < 1$,

$$\int_0^{\infty} \frac{\sin s}{s} \cos(s\rho \cos \beta) \, \mathrm{d}s = \pi/2 \qquad \text{(A7-14)}$$

$$\int_0^{\infty} \cos(s) \cos(s\rho \cos \beta) \, \mathrm{d}s = 0 \qquad \text{(A7-15)}$$

Accordingly, within the inclusion, the integral Q_{ij} is given by:

$$Q_{ij}(\rho < 1) = \frac{K_{kl}}{4\pi} \int_0^{2\pi} \int_0^{\pi} \bar{G}^0_{ijkl}(\varphi, \eta) \sin \eta \, \mathrm{d}\eta \, \mathrm{d}\varphi \qquad \text{(A7-16)}$$

In view of eqn. (147), this is simply

$$Q_{ij}(\rho < 1) = E^0_{ijkl} K_{kl} \qquad \text{(A7-17)}$$

Substituting eqn. (A7-17) into (A7-3) and noting that R_{ijkl} is constant within the inclusion reveals that p^2_{ij} should be a constant within the inclusion and zero elsewhere. This implies that the stress polarisation field is indeed constant within the inclusion and is given by the solution of the following algebraic equation:

$$p_{ij} = R_{ijkl}(\varepsilon^{\infty}_{kl} + E^0_{klmn} p_{mn}) \qquad \text{(A7-18)}$$

where E^0_{ijkl} is given by eqn. (147). The strain field within the inclusion is readily obtained as:

$$\varepsilon_{ij} = \varepsilon^{\infty}_{ij} + E^0_{ijkl} p_{kl} \qquad \text{(A7-19)}$$

which is also constant. It should be noted, however, that the strain outside the inclusion is not uniform and can be obtained by interpreting $Q_{ij}(\mathbf{r})$ as ε'_{ij}, the deviation of the strain field from $\varepsilon^{\infty}_{ij}$, and integrating eqn. (A7-13) for $\rho > 1$.

Chapter 8

STRESS CONCENTRATIONS AROUND CIRCULAR HOLES IN A FILAMENT STIFFENED SHEET: SOME EXPERIMENTAL RESULTS

D. J. MALCOLM†
Lakehead University, Ontario, Canada

SUMMARY

The purpose of the results contained herein is to give experimental confirmation of theoretical predictions[2,6] and to support observed behaviour of filamentary composites under fatigue loading[10] when higher strengths were recorded than static tests would have predicted. This reflects on the stress distribution around the tip of a fatigue crack the radius of which is comparable to the diameter of the filament.

The present results support those of Ref. 2 and others in that the stress concentration is indeed reduced if only a few fibres are cut. However, their assumptions lead to an overestimate of this reduction.

Further experimental work should investigate models having different geometries and different materials. Attention should also be directed towards the behaviour at inclusions near ultimate loads.

INTRODUCTION

The use of any material is governed largely by its strength at connections, irregularities and other points of stress concentration. When viewed as homogeneous and isotropic much work has been accomplished in creating mathematical models and solutions to the stress distribution around holes of various shapes and subject to several loading patterns.[7-9] Further, the

---

† *Present address:* The University of Calgary, Canada.

similar behaviour in homogeneous but orthotropic materials has been well documented.[7,8]

Composite materials have commonly been viewed as orthotropic and homogeneous[5] and this approach is valid on a macroscopic level. On a microscopic level, however, the heterogeneity is of importance and if the size of an inclusion or irregularity is comparable to the dimension of the internal structuring then the interaction between matrix and fibres or laminates must be considered.

Such an approach was first taken by Hedgepeth[4] and developed further by Fichter,[1] Franklin[2] and Kulkarni et al.,[6] all of whom used mathematical models of idealised materials to demonstrate that as the number of fibres cut by a hole within a uniform stress field is reduced, then so is the stress concentration factor also reduced.

Experimental confirmation of these results has been limited to the explanation of the observed high fracture toughness of some advanced composites;[10] but a direct confirmation of the stress pattern around such holes has not appeared in the literature to the writer's knowledge.

The purpose of the present paper is to partially fill this need for experimental confirmation as has been suggested in the literature.[2] The following sections therefore describe the construction of an enlarged model of a layered material with two sizes of circular holes, its testing and the stresses recorded. These results are compared with those predicted in Ref. 2. Comparison is also made with closed form and finite element results for an equivalent orthotropic and homogeneous material.

EXPERIMENTAL STUDY

The intention of the tests was to approximate the proportions and properties of a high-strength fibre composite on a sufficiently large scale to enable strain measurements to be taken both on the matrix and on the fibres. Furthermore it was intended to examine only the effect of a circular hole within a uniform tension field.

The model constructed is shown in Fig. 1. The materials used were polyvinylchloride (PVC) and aluminium alloy the elastic moduli of which were in the approximate ratio of 1:21, thus simulating the composition of a typical glass fibre–epoxy composite.

The two components were joined with Loctite's Superbonder 150 contact adhesive. Despite the requirement that the bonding surfaces be in close contact with each other and despite the brittleness of the bond this adhesive

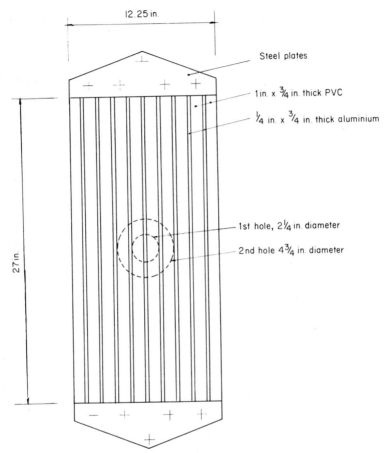

FIG. 1.   Details of composite test model.

was chosen because of its low ductility which ensured minimal slippage. When testing samples it offered a shear strength of 800 psi ($55 \cdot 2 \times 10^6$ Pa). Other adhesives considered and/or tested were a range of epoxy cements, Eastman 910 and other adhesives in the Superbonder series.

Loading was applied to the fishtail plate at each end and transmitted to the composite through four bolts and steel strips which were bonded to each side of the composite. Preliminary testing with loads up to 10 kips (44·5 kN) indicated a uniform strain field across the centre of the composite.

The next stage involved the cutting of a circular hole at the centre of the plate. This hole cut only the central aluminium fibre and was tangential to

D. J. MALCOLM

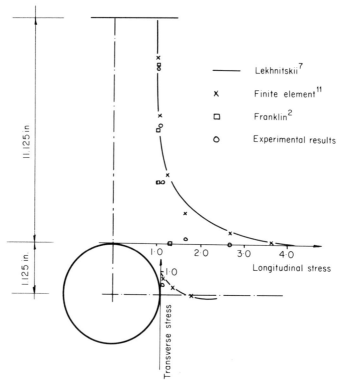

FIG. 2.   Longitudinal and transverse stress distributions: 2·25 in diameter hole.

the adjacent fibres. The second hole was formed by enlarging the first and cut three fibres and was also tangential to the adjacent fibres. These holes are shown in broken lines in Fig. 1.

The ratio of hole diameter to overall width of the model was 1:5·44 for the smaller hole and 1:2·58 for the larger hole. According to Ref. 8 the size of the smaller hole would therefore approximate a hole in an infinite medium.

Epoxy-backed electrical strain gauges were affixed to both the PVC and the aluminium fibres, some being placed on the inside circumference of the hole and others on the outer surface of the uncut fibres. Strain measurements were taken using a Vishay multi-channel recorder; the only problem encountered was the tendency of the resistance of gauges attached to the PVC to vary as current passed through them.

Stresses recorded on the unbroken fibres at the centre section are shown in Figs. 2 and 3, together with transverse stresses on the centre line adjacent

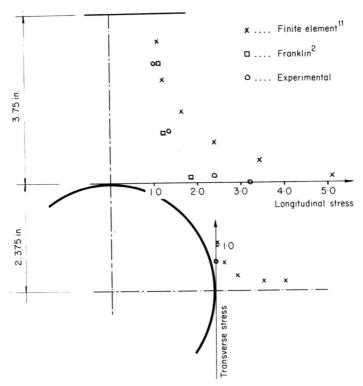

FIG. 3.   Longitudinal and transverse stress distributions: 4·75 in diameter hole.

to the holes. These results were adjusted to give a unit longitudinal stress in the outermost aluminium fibre.

Results from stress–strain tests on tension samples of the aluminium and PVC indicated elastic moduli of $8·33 \times 10^6$ psi ($5·75 \times 10^{10}$ Pa) and $4·04 \times 10^5$ psi ($2·79 \times 10^9$ Pa), respectively, and Poisson ratios of $0·342$ and $0·372$, respectively.

THEORETICAL RESULTS

Viewed as an orthotropic material the equivalent three (independent) elastic constants may be derived from a strength-of-materials approach.[5] For the specimen shown in Fig. 1 having the component elastic properties indicated above, the following orthotropic properties are obtained: E_x (longitudinal)

$= 2 \cdot 032 \times 10^6 \, \text{psi}$ ($1 \cdot 40 \times 10^{10} \, \text{Pa}$); E_y (transverse) $= 0 \cdot 587 \times 10^6 \, \text{psi}$ ($0 \cdot 405 \times 10^{10} \, \text{Pa}$); v_{xy} (transverse strain caused by unit longitudinal strain) $= 0 \cdot 32$.

Using these values in an expression quoted in Ref. 7 the maximum longitudinal stress was calculated (assuming an infinite medium) and is shown in Fig. 2.

A second prediction based on the equivalent orthotropic properties utilised a plane stress constant strain triangle finite element programme.[11] Some of the results from this analysis are also shown in Figs. 2 and 3.

Instead of viewing the composite as an equivalent orthotropic one, Franklin[2] and Kulkarni et al.[6] have derived exact stress distributions in an infinite medium based on the idealisation that the matrix carries no longitudinal stress and that the fibres are not subject to shear deformation (i.e. the fibres are extremely thin). Both papers indicate a maximum fibre stress concentration factor of 1·33 when only one fibre has been cut, and although Kulkarni's work claims more correct boundary conditions both give very similar stress concentration factors when more than one fibre is cut.

The results for longitudinal fibre stresses at critical points based on Ref. 2 are also shown in Figs. 2 and 3. These results and those based upon Ref. 7 correspond to a unit stress applied to the infinite medium.

DISCUSSION

Figure 2 shows the considerable difference between the longitudinal stresses calculated on the basis of a homogeneous material[7] and the distribution obtained on the basis of an idealised filamentary composite[2] in which only one fibre is cut. Not unexpectedly the experimental results fall between these two theoretical extremes, although if the average stress in the first uncut fibre is considered then a maximum stress concentration of approximately 1·8 is obtained from the experiment; this is closer to the filamentary results (1·33) than to the homogeneous model's prediction of 4·2. Indeed one assumption in Ref. 2 is that the fibres are very thin or have a uniform stress.

Both Refs. 2 and 7 assumed an infinite medium, whereas the experimental model has an overall width to (small) hole diameter ratio of 5·44. However, the finite element results (based on a unit stress applied to the gross section) agree closely with Ref. 7. Therefore, it is unlikely that experimental results would have differed much if a wider plate had been used.

However, the second set of experimental results, shown in Fig. 3, from a hole which cut three fibres and for which the width to the diameter ratio was 2·58, cannot be compared strictly with Refs. 2 or 7. Nevertheless it is of interest to note that the maximum stress (3·17) is considerably less than the finite element analysis prediction for a homogeneous material of near 5·5.

ACKNOWLEDGEMENTS

The experimental results presented here are based on those of Mr B. Staus whose patience and industry are gratefully recognised. The writer also wishes to acknowledge the financial support of the National Research Council of Canada through grant A9316 and of the President's NRC fund of Lakehead University.

REFERENCES

1. FICHTER, W. B. Stress Concentrations around Broken Filaments in a Filament Stiffened Sheet, NASA TN D-5433, 1969.
2. FRANKLIN, H. G. Hole stress concentrations in filamentary structures, *Fibre Sci. Technol.*, **2** (1970), pp. 241–9.
3. GRESZCZUK, L. B. Stress Concentrations and Failure Criteria for Orthotropic and Anisotropic Plates with Circular Openings, Composite Materials: Testing and Design (2nd Conf.), ASTM STP 497 (1972), pp. 363–81.
4. HEDGEPETH, J. M. Stress concentration in filamentary structures, *Fibre Sci. Technol.*, **2** (1970), p. 240.
5. JONES, R. M. *Mechanics of Composite Materials*, McGraw-Hill, New York, 1975.
6. KULKARNI, S. V., ROSEN, B. W. and ZWEBEN, C. Load concentration factors for circular holes in composite laminates, *J. Composite Materials*, **7** (July 1974), pp. 387–93.
7. LEKHNITSKII, S. G. *Theory of Elasticity of an Anisotropic Elastic Body*, Holden-Day, San Francisco, 1963.
8. SAVIN, G. N. *Stress Concentrations around Holes*, Pergamon, Oxford, 1961.
9. TIMOSHENKO, S. and GOODIER, J. N. *Theory of Elasticity*, McGraw-Hill, New York, 1970.
10. WADDOUPS, M. E. *et al.* Macroscopic fracture mechanics of advanced composite materials, *J. Composite Materials*, **5** (1971), p. 446.
11. ZIENKIEWICZ, O. C. *The Finite Element Method in Engineering Science*, McGraw-Hill, London, 1971.

Chapter 9

STRESSES IN THREE-DIMENSIONAL COMPOSITES WITH LIMITING SHEAR PROPERTIES

P. S. Theocaris & S. A. Paipetis

National Technical University of Athens, Greece

SUMMARY

A particular kind of particle reinforcement results when the filler shear modulus is equal to that of the matrix, but the respective Poisson's ratio is higher. In this case, slightly higher values of Young's modulus of the composite appear, but considerably higher values of the bulk modulus. On the other hand, high stress gradients disappear from the stress fields around the filler particle, and reinforcement occurs under conditions of minimal stress concentration, the structural integrity of the composite being thus preserved.

INTRODUCTION

Reinforced composites are produced by means of a hard filler, which is embedded in a soft matrix. In particular, by 'reinforcement' one means a procedure by which a number of parameters, corresponding to certain mechanical properties of a material, may acquire higher values. Such properties may be the mechanical moduli, the tensile or compressive strength, the fatigue limit, etc., of the material. Introduction of a filler into a matrix material generally affects most of its mechanical properties, because of the stress/strain fields created around the filler particles. The stronger the latter fields, the higher is the reinforcement.

However, points of high stress concentration can always be the origin of fatigue cracks or even brittle fracture phenomena, and an optimum solution between the two contradicting situations is required. In particular, when one is interested in obtaining higher Young's or bulk modulus for the composite, while no higher shear modulus is required, constituent materials with limiting shear properties, i.e. nearly equal shear moduli, can be applied.

The stress field created around a filler inclusion under the action of a uniform stress at infinity generally decays with the distance from the centre of the inclusion. The rate of decay depends on the matrix-to-filler shear moduli ratio. When this becomes equal to unity, all terms in the expressions for the stress, containing high powers of the inverse of the distance, vanish.

In this manner, no considerable stress gradients develop in the composite, while on the other hand the values of Young's or bulk modulus depend on the relation of Poisson's ratios of matrix and filler.

The plane problem was considered in a recent paper by the authors.[1] The stress field developed around a circular elastic inclusion, embedded in an infinitely extended elastic matrix under the action of a tensile stress at infinity, was investigated for the case of equal shear moduli for the matrix and the inclusion, under conditions of both plain strain and generalised plane stress. It was proved that, in the expressions for the stresses, only terms containing squares of the inverse distance were retained, while terms containing fourth powers of the same disappeared. In addition it was shown that, in this way, a composite of the same shear modulus can be produced, as its constituents, but with higher Young's modulus, if Poisson's ratio of the filler exceeds that of the matrix, and this material works under ideal conditions as far as its structural integrity is concerned.

However, the plane model so far considered is not sufficient to simulate a fully three-dimensional composite, as it is the case of particle reinforcement. Therefore, the problem of the stress fields developed in the case of an elastic sphere, embedded in an infinitely extended elastic matrix, under the action of a tensile stress p at infinity, appears to be of interest.

THEORY

The stress components in and around the inclusion are given in spherical coordinates r, φ, ϑ as follows[2] (index m refers to matrix, index f to inclusion):

$$\sigma_{rr}^{f} = p\left[\frac{\lambda_m + 2G_m}{3\lambda_m + 2G_m}\frac{3\lambda_f + 2G_f}{3\lambda + 2G_f + 4G_m} + 4G_f\frac{D_1}{D}P_2(\cos\vartheta)\right]$$

$$\sigma_{\vartheta\vartheta}^{f} = p\left[\frac{\lambda_m + 2G_m}{3\lambda_m + 2G_m}\frac{3\lambda_f + 2G_f}{3\lambda_f + 2G_f + 4G_m} + 2G_f\frac{D_1}{D} - 4G_f\frac{D_1}{D}P_2(\cos\vartheta)\right]$$

$$\sigma_{\varphi\varphi}^{f} = p\left[\frac{\lambda_m + 2G_m}{3\lambda_m + 2G_m}\frac{3\lambda_f + 2G_f}{3\lambda_f + 2G_f + 4G_m} - 2G_f\frac{D_1}{D}\right] \tag{1}$$

$$\sigma_{r\vartheta}^{f} = p\left[2G_{f}\frac{D_{1}}{D}\frac{dP_{2}(\cos\vartheta)}{d\vartheta}\right]$$

$$\sigma_{rr}^{m} = p\left[\frac{1}{3} + \frac{4G_{m}}{3\lambda_{m} + 2G_{m}}\frac{3(\lambda_{f} - \lambda_{m}) + 2(G_{m} - G_{f})}{9\lambda_{f} + 6G_{f} + 12G_{m}}\frac{R^{3}}{r^{3}}\right.$$
$$\left. + \left(\frac{2}{3} + \frac{9\lambda_{m} + 10G_{m}}{3}\frac{D_{2}}{D}\frac{R^{5}}{r^{5}}\right)P_{2}(\cos\vartheta)\right]$$

$$\sigma_{\varphi\varphi}^{m} = p\left[\frac{2}{3} - \left(\frac{2G_{m}}{3\lambda_{m} + 2G_{m}}\frac{3(\lambda_{f} - \lambda_{m}) + 2(G_{m} - G_{f})}{9\lambda_{f} + 6G_{f} + 12G_{m}} + \frac{G_{m}}{3}\frac{D_{2}}{D}\right)\frac{R^{3}}{r^{3}}\right.$$
$$\left. + 2G_{m}\frac{D_{3}}{D}\frac{R^{5}}{r^{5}} - \left(\frac{2}{3} + \frac{G_{m}}{3}\frac{D_{2}}{D}\frac{R^{3}}{r^{3}} + 14G_{m}\frac{D_{3}}{D}\frac{R^{5}}{r^{5}}\right)P_{2}(\cos\vartheta)\right]$$

$$\sigma_{\varphi\varphi}^{m} = p\left[\left(\frac{G_{m}}{3}\frac{D_{2}}{D} - \frac{2G_{m}}{3\lambda_{m} + G_{m}}\frac{3(\lambda_{f} - \lambda_{m}) + 2(G_{f} - G_{m})}{9\lambda_{f} + 6G_{f} + 12G_{m}}\right)\frac{R^{3}}{r^{3}} - 2G_{m}\frac{D_{3}}{D}\frac{R^{5}}{r^{5}}\right.$$
$$\left. - \left(G_{m}\frac{D_{2}}{D}\frac{R^{3}}{r^{3}} + 10G_{m}\frac{D_{3}}{D}\frac{R^{5}}{r^{5}}\right)P_{2}(\cos\vartheta)\right] \qquad (2)$$

$$\sigma_{r\vartheta}^{m} = p\left[\left(\frac{1}{3} - \frac{3\lambda_{m} + 2G_{m}}{6}\frac{D_{2}}{D}\frac{R^{3}}{r} - 8G_{m}\frac{D_{3}}{D}\frac{R^{5}}{r^{5}}\right)\frac{dP_{2}(\cos\vartheta)}{d\vartheta}\right]$$

These expressions have been derived under the assumption of perfect adhesion, i.e. continuity of stresses and displacements at the interface, and they contain λ, G, the Lamé constants of the constituent materials, R the radius of the inclusion, while by $P_{2}(\cos\vartheta)$ and $dP_{2}(\cos\vartheta)/d\vartheta$ a Legendre polynomial and its associated Legendre polynomial are denoted. Further it is valid that:

$$D = \begin{vmatrix} \lambda_{f}/7 & 4G_{f} & -\dfrac{9\lambda_{m} + 10G_{m}}{3} & -24G_{m} \\[2ex] -\dfrac{8\lambda_{f} + 7G_{f}}{21} & 2G_{f} & \dfrac{3\lambda_{m} + 2G_{m}}{6} & 8G_{m} \\[2ex] -\lambda_{f}/7 & 2 & \dfrac{3\lambda_{m} + 5G_{m}}{6G_{m}} & 3 \\[2ex] -\dfrac{5\lambda_{f} + 7G_{f}}{42G_{f}} & 1 & 1/6 & -1 \end{vmatrix} \qquad (3)$$

$$D_1 = \begin{vmatrix} \lambda_f/7 & 2/3 & -\dfrac{9\lambda_m + 10G_m}{3} & -24G_m \\[2em] -\dfrac{8\lambda_f + 7G_f}{21} & 1/3 & \dfrac{3\lambda_m + 2G_m}{6} & 8G_m \\[2em] -\lambda_f/7 & 1/3G_m & \dfrac{3\lambda_m + 5G_m}{6G_m} & 3 \\[2em] -\dfrac{5\lambda_f + 7G_f}{42G_f} & 1/6G_m & 1/6 & -1 \end{vmatrix} \qquad (4)$$

$$D_2 = \begin{vmatrix} \lambda_f/7 & 4G_f & 2/3 & -24G_m \\[2em] -\dfrac{8\lambda_f + 7G_f}{21} & 2G_f & 1/3 & 8G_m \\[2em] -\lambda_f/7 & 2 & 1/3G_m & 3 \\[2em] -\dfrac{5\lambda_f + 7G_f}{42G_f} & 1 & 1/6G_m & -1 \end{vmatrix} \qquad (5)$$

$$D_3 = \begin{vmatrix} \lambda_f/7 & 4G_f & -\dfrac{9\lambda_m + 10G_m}{3} & -24G_m \\[2em] -\dfrac{8\lambda_f + 7G_f}{21} & 2G_f & \dfrac{3\lambda_m + 2G_m}{6} & 8G_m \\[2em] -\lambda_f/7 & 2 & \dfrac{3\lambda_m + 5G_m}{6G_m} & 3 \\[2em] -\dfrac{5\lambda_f + 7G_f}{42G_f} & 1 & 1/6 & -1 \end{vmatrix} \qquad (6)$$

For $G_f = G_m$, one obtains $D = 6G_m D_1$ and $D_2 = D_3 = 0$, and the above expressions become:

$$\frac{\sigma_{rr}^f}{p} = \frac{1}{3}\frac{1-v_m}{1+v_m}\frac{1+v_f}{1-v_f} + \frac{2}{3}P_2(\cos\vartheta)$$

$$\frac{\sigma_{\vartheta\vartheta}^f}{p} = \frac{1}{3}\frac{1-v_m}{1+v_m}\frac{1+v_f}{1-v_f} + \frac{1}{3} - \frac{2}{3}P_2(\cos\vartheta)$$

$$\frac{\sigma_{\varphi\varphi}^f}{p} = \frac{1}{3}\frac{1-v_m}{1+v_m}\frac{1+v_f}{1-v_f} - \frac{1}{3}$$

$$\frac{\sigma_{r\vartheta}^f}{p} = \frac{1}{3}\frac{\mathrm{d}P_2(\cos\vartheta)}{\mathrm{d}\vartheta}$$

$$(7)$$

$$\frac{\sigma_{rr}^m}{p} = \frac{1}{3} + \frac{2}{9}\frac{v_f - v_m}{(1+v_m)(1-v_f)}\frac{R^3}{r^3} + \frac{2}{3}P_2(\cos\vartheta)$$

$$\frac{\sigma_{\vartheta\vartheta}^m}{p} = \frac{2}{3} - \frac{1}{9}\frac{v_f - v_m}{(1+v_m)(1-v_f)}\frac{R^3}{r^3} - \frac{2}{3}P_2(\cos\vartheta)$$

$$\frac{\sigma_{\varphi\varphi}^m}{p} = -\frac{1}{9}\frac{v_f - v_m}{(1+v_m)(1-v_f)}\frac{R^3}{r^3}$$

$$\frac{\sigma_{r\vartheta}^m}{p} = \frac{1}{3}\frac{\mathrm{d}P_2(\cos\vartheta)}{\mathrm{d}\vartheta}$$

$$(8)$$

In the last expressions one may notice the following:

(1) A complete separation of the variables has taken place. Namely, there are constant terms, terms which are functions of the ratio R/r and terms which depend on the azimuthal angle ϑ only.

(2) In the inclusion, as expected, stresses do not depend on the radial distance r.

(3) In the matrix, terms representing high stress gradients, depending on R^5/r^5, have disappeared.

(4) The sum of principal stresses in the matrix is equal to the applied stress at infinity p, while in the inclusion depends on both matrix and inclusion Poisson's ratios. This is the case with the plane problem also.[1]

Stresses σ_{rr} and $\sigma_{r\vartheta}$ both in matrix and the inclusion depend on a term $2/3 P_2(\cos\vartheta)$, while shear stress $\sigma_{r\vartheta}$ on term $(1/3)(\mathrm{d}P_2(\cos\vartheta)/\mathrm{d}\vartheta)$. Both terms,

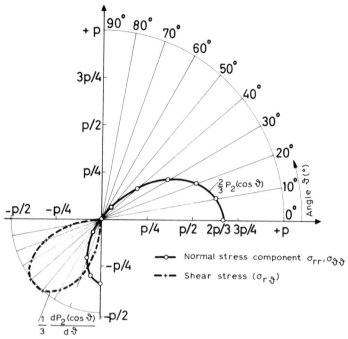

FIG. 1.   Normal and shear stress distribution around a spherical elastic inclusion embedded in an infinitely extended elastic matrix of equal shear modulus under the action of a tensile stress p at infinity.

i.e. the azimuthal-angle-dependent components of the stresses, are plotted in Fig. 1.

Normal stresses in the inclusion depend on the expression

$$p\left(\frac{1 - v_m}{1 + v_m}\right)\left(\frac{1 + v_f}{1 - v_f}\right)$$

which is equal to the sum A_f of the latter (one may compare with the respective value $p(1 - v_m)/(1 - v_f)$ for plane strain and $p(1 + v_f) \times (1 - 0.75 v_m)/[(1 + v_m)(1 - 0.75 v_f)]$ for generalised plane stress). The ratio A_f/A_m plotted against v_m, for various v_f, is presented in Fig. 2.

DISCUSSION

Let a two-phase composite now be considered whose $G_m = G_f$, but $v_m \neq v_f$. Several theories exist for the prediction of its Young's or bulk modulus. The

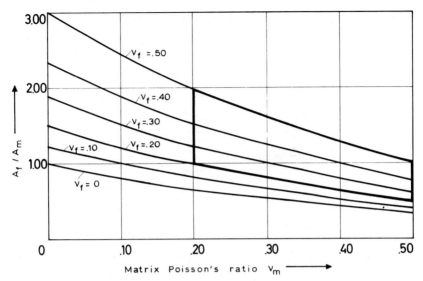

FIG. 2. Ratio of the sum of principal stresses in the matrix and the inclusions for various Poisson's ratio combinations.

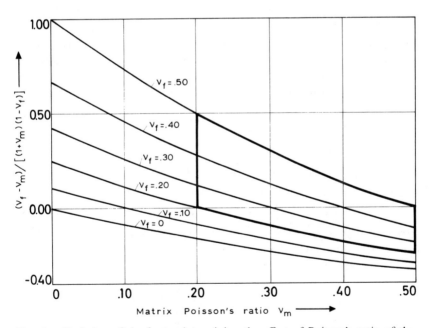

FIG. 3. Variation of the factor determining the effect of Poisson's ratio of the constituents on the stress distribution in the matrix.

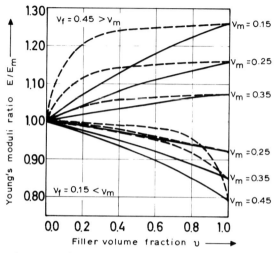

FIG. 4.   Composite-to-matrix E: moduli ratio based on various theories. ———:
Hashin;[6] ————: Kerner.[5]

law of mixtures in its linear or non-linear form, or Paul's upper and lower
bounds,[3] is only a highly qualitative measure in the present case, as it is valid
for equal Poisson's ratios. Effective bounds based on variational principles,
see for example Ref. 4, are expected to provide a more or less accurate
approach, as they close together for $G_f \rightarrow G_m$. However, direct methods, i.e.
based on the evaluation of stress fields developed around individual
inclusions, are of immediate interest. Concerning Young's modulus E and
bulk modulus K, the following expressions have been given. According to
Kerner:[5]

$$K = \frac{\dfrac{K_m(1-v)}{3K_m + 4G_m} + \dfrac{K_f v}{3K + 4G_m}}{\dfrac{1-v}{3K_m + 4G_m} + \dfrac{v}{3K_f + 4G_m}}$$

from which K/K_m and E/E_m can be obtained in terms of v_m and v_f for G_m
$= G_f$. The volume fraction of the filler is denoted by v.
   The respective expression given by Hashin[6] is:

$$\frac{K}{K_1} = 1 + \frac{3(1 - v_m)\left(\dfrac{K_f}{K_m} - 1\right)v}{2(1 - 1 v_m) + (1 + v_m)\left[\dfrac{K_f}{K_m} - \left(\dfrac{K_f}{K_m} - 1\right)v\right]}$$

In Fig. 4, Young's moduli ratio E/E_m is presented for both Hashin and Kerner theories. According to the first of them, which appears to be closer to reality, for $v = 0.30$ a 10% increase of Young's modulus is expected for $v_m = 0.15$ and $v_f = 0.45$. On the contrary, for interchanged values of Poisson's ratios, a drop equal to 6.5% occurs.

In Fig. 5 the bulk moduli ratio is given, again for both theories of Refs. 5 and 6. Here, for $G_f = G_m$, the increase of bulk modulus appears to be much more considerable. For example, for $v = 0.30$, $v_m = 0.15$, $v_f = 0.45$, it is equal to 70% according to Hashin. For $v_m > v_f$, bulk modulus decreases, but at a slower rate with $(v_m - v_f)$.

Let a real case now be considered. A diglycidyl ether of bisphenol A resin of the Epikote 828 type (Shell Co.), cured by an 8% amount of triethylene tetramine, can have its mechanical properties modified by means of a

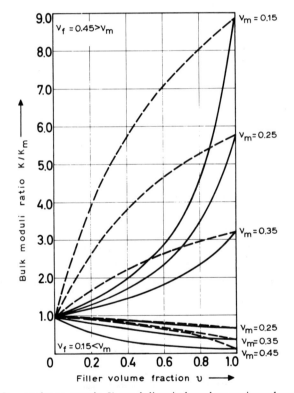

FIG. 5. Composite-to-matrix K: moduli ratio based on various theories. ———: Hashin;[6] ———: Kerner.[5]

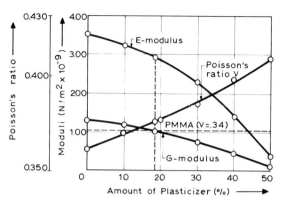

FIG. 6.   Mechanical properties of an amine-cured cold-setting epoxy resin with various degrees of plastification as compared with the respective properties of a polymethylmethacrylate.

suitable plasticising agent, such as a polysulphide of the Thiokol LP-3 type. The variation of E and G moduli, as well as of Poisson's ratio v, with the amount of plasticiser by weight, appears in Fig. 6. If a combination with another type of polymer is considered, for example a polymethylmethacrylate (PMMA), one obtains equal shear moduli at a plasticiser amount equal to 18 % for the epoxy resin. The respective values of Poisson's ratios are 0·34 for PMMA and 0·372 for the resin.

A PMMA matrix filled with 50 % by volume with particles of the 10 % plasticised resin, for this small difference ($v_f - v_m$) = 0·032 only exhibits 1·22 % increase according to Hashin and 1·38 % according to Kerner. But, for the bulk modulus, these values become 12·78 % and 14·63 %. For 80 % filler volume fraction, the respective values for Young's modulus are 1·93 % and 2·02 % and for bulk modulus 21·57 % and 22·79 %.

However, for higher values of ($v_f - v_m$), more considerable increase can be obtained, which, when combined with other important properties of the materials, such as strength or vibration damping, may lead to interesting results.

REFERENCES

1. THEOCARIS, P. S. and PAIPETIS, S. A. *J. Comp. Mat.*, **9** (1975), No. 3, p. 244.
2. SEZAWA, K. and MIYAZAKI, B. *Proc. Jap. Soc. Mech. Eng.*, **31** (1928), No. 136, p. 625.

3. PAUL, B. *Trans. Met. Soc. AIME*, **218** (1960), p. 36.
4. HASHIN, Z. and SHTRIKMAN, S. *J. Mech. Phys. Solids*, **10** (1962), p. 335.
5. KERNER, E. H. *Proc. Phys. Soc.*, **69B** (1956), p. 808.
6. HASHIN, Z. *J. Appl. Mech.*, **84** (1962), p. 143.

Chapter 10

A PLANE STRAIN SOLUTION OF AN ELASTIC CYLINDRICAL INCLUSION IN AN ELASTO-PLASTIC MATRIX UNDER UNIFORM TENSION

D. K. Brown

The University of Glasgow, Scotland, UK

SUMMARY

An incremental solution is presented for the plane strain problem of an elastic cylindrical inclusion in an elasto-plastic matrix under the action of uniform tension applied at finite and infinite outer boundaries. Results are given in the form of stress and shear strain concentration ratios and the development of maximum interfacial stress and shear strain with load.

NOMENCLATURE

a	Radius of inclusion.
C_1, C_2	Constants of integration.
E, E'	Modulus of elasticity of matrix, inclusion.
EPO	Effective plastic strain remote from inclusion.
Q	Work-hardening index.
r	Radius.
u	Radial displacement.
R	Radius of finite outer boundary.
T	Geometric ratio of radii.
$d\bar{\varepsilon}^P, d\varepsilon_r^P, d\varepsilon_\theta^P, d\varepsilon_z^P$	Current increments of effective and component strains.
$\varepsilon_r^P, \varepsilon_\theta^P, \varepsilon_z^P$	Accumulated plastic strains.
$\Sigma_r, \Sigma_\theta, \Sigma_z$	Total plastic strains.
v	Poisson's ratio.

$\sigma_r, \sigma_\theta, \sigma_z$	Normal stresses.
$\bar{\sigma}$	Effective stresses.
σ_Y	Yield stress.
$\alpha_1, \alpha_2, \alpha_3, \delta$	Functions of Poisson's ratio.
$\beta_1, \beta_2, \beta_R$	Functions of Poisson's ratio and elastic moduli.
$I_1^i, I_2^i, DI_1^i, DI_2^i$	Definite integrals of plastic strains with limits 1 to i.

INTRODUCTION

In 1974, Orr and Brown[1] published a solution to the plane strain problem of a rigid cylindrical inclusion in an elasto-plastic matrix under various states of stress. The case of uniform applied stress was not considered, nor was there any elasticity allowed in the inclusion. Some features of the solution required confirmation, such as Figs. 5 and 6 of that paper. Worry was expressed by Argon et al.[2] about these unusual features.

The solution presented below is once again the cylindrical inclusion embedded in an elasto-plastic matrix, except that in this case elasticity of the inclusion is allowed and the applied loading is a uniform stress (Fig. 1). The problem, being symmetrical in geometry and loading, is one dimensional, thus reducing the complexity of solution. The approach of Orr and Brown was to expand the two governing simultaneous second-order partial differential equations of the two displacements into finite differences and solve the boundary value problem using successive over-relaxation. The approach in the present solution is quite different from the above and is similar to that used by Davis[3] to solve the elasto-plastic problem of a circular hole in a plate in uniform tension. However, unlike Davis, instead of a hole there is an elastic inclusion and the technique of component plastic strain determination is that of Mendelson.[4,5] The problem is reduced to a second-order differential equation in the total circumferential strain ε_θ, which, after integration of the equation, can be expressed in terms of integrals of plastic strains and two constants. The constants are found by using boundary conditions at the inclusion/matrix interface and the outer load boundary.

Both inclusion and matrix have the same Poisson's ratio, 0·3, but different moduli of elasticity, E' and E, respectively. The ratio E/E' thus describes the relative rigidity of matrix to inclusion, it being zero for a rigid inclusion and infinitely large when the inclusion is replaced by a hole. Four values of ratio were used: 0·0, 0·5, 1·0, 10·0; the value of 0·5 is similar to the matrix/particle modulus ratio, which might be expected from alumina

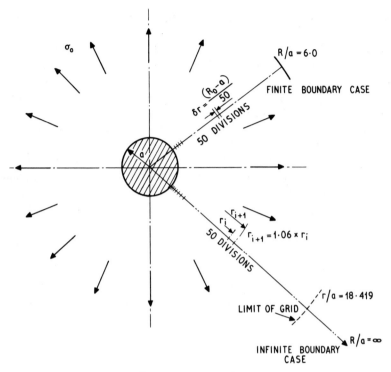

FIG. 1. Details of grids.

particles in copper. The value of 10·0 on the other hand represents the case of very soft inclusions in an elasto-plastic matrix. In this latter case, it would seem unlikely that the inclusion would remain elastic, but in any case should the inclusion begin to flow plastically it would become equivalent to an even softer inclusion.

At all times the inclusion is assumed elastic. The matrix is assumed to linearly work-harden and two values of coefficient Q were used, 0·0 and 0·025, where

$$Q = \frac{1}{E} [\mathrm{d}\bar{\sigma}/\!\int \mathrm{d}\bar{\varepsilon}^P]$$

and

$$\bar{\sigma} = \{\tfrac{1}{2}[(\sigma_r - \sigma_\theta)^2 + (\sigma_\theta - \sigma_z)^2 + (\sigma_z - \sigma_r)^2]\}^{1\,2}$$

$$\mathrm{d}\bar{\varepsilon}^P = \{\tfrac{2}{9}[(\mathrm{d}\varepsilon_r^P - \mathrm{d}\varepsilon_\theta^P)^2 + (\mathrm{d}\varepsilon_\theta^P - \mathrm{d}\varepsilon_z^P)^2 + (\mathrm{d}\varepsilon_z^P - \mathrm{d}\varepsilon_r^P)^2]\}^{1\,2}$$

the component shear stresses and strains being zero in this problem due to symmetry.

It is of interest to note that in 1968, Mendelson[6] tackled a similar problem, in which he used the deformation theory of plasticity and maximum applied loads of up to only twice the initial yield load. In the approach adopted here the more accurate incremental theory is used with applied loads up to 32 times the initial yield load. Mendelson solved for only the infinite outer boundary case and used a Poisson's ratio of 0·35 throughout, whereas here a finite outer boundary is also considered and Poisson's ratio is 0·3. The results presented in Ref. 6 are in the form of effective stress and effective total strain concentration factors and only one intermediate case of E/E' ($= 0·02$) is considered.

ANALYSIS

All stresses, strains and radii are non-dimensional: $\sigma \equiv \sigma/\sigma_Y$, $\varepsilon \equiv E\varepsilon/\sigma_Y$, $r \equiv r/a$. Under plane strain conditions, $\varepsilon_z = 0$. In cylindrical polar co-ordinates equilibrium is given as

$$\sigma_\theta = r \frac{d\sigma_r}{dr} + \sigma_r \tag{1}$$

The elastic strains can always be determined from the stresses or as a component of total strain

$$\varepsilon_r^e = [\sigma_r - v(\sigma_\theta + \sigma_z)] = \varepsilon_r - \varepsilon_r^P - d\varepsilon_r^P \tag{2a}$$

$$\varepsilon_\theta^e = [\sigma_\theta - v(\sigma_r + \sigma_z)] = \varepsilon_\theta - \varepsilon_\theta^P - d\varepsilon_\theta^P \tag{2b}$$

$$\varepsilon_z^e = [\sigma_z - v(\sigma_\theta + \sigma_r)] = \varepsilon_z - \varepsilon_z^P - d\varepsilon_z^P$$

$$= 0 + (\varepsilon_\theta^P + \varepsilon_r^P) + (d\varepsilon_\theta^P + d\varepsilon_r^P) \tag{2c}$$

Use is made in eqn. (2c) of the constant volume plastic condition. Thus

$$\sigma_z = v(\sigma_r + \sigma_\theta) + (\varepsilon_\theta^P + d\varepsilon_\theta^P) + (\varepsilon_r^P + d\varepsilon_r^P) \tag{3}$$

Substitution of (3) into (2a) and (2b) gives

$$\varepsilon_\theta = (1 - v^2)\sigma_\theta - v(1 + v)\sigma_r + (1 - v)(\varepsilon_\theta^P + d\varepsilon_\theta^P) - v(\varepsilon_r^P + d\varepsilon_r^P)$$

$$\varepsilon_r = (1 - v^2)\sigma_r - v(1 + v)\sigma_\theta + (1 - v)(\varepsilon_r^P + d\varepsilon_r^P) - v(\varepsilon_\theta^P + d\varepsilon_\theta^P) \tag{4}$$

Elimination of σ_0 from (4) gives

$$v\varepsilon_0 + (1 - v)\varepsilon_r = (1 + v)(1 - 2v)\sigma_r + (1 - v)(\varepsilon_r^P + d\varepsilon_r^P)$$

or

$$\sigma_r = \frac{1 - v}{(1 + v)(1 - 2v)}\varepsilon_r + \frac{v}{(1 + v)(1 - 2v)}\varepsilon_0 - \frac{1}{(1 + v)}(\varepsilon_r^P + d\varepsilon_r^P)$$

$$\equiv \alpha_1\varepsilon_r + \alpha_2\varepsilon_0 - \alpha_3(\varepsilon_r^P + d\varepsilon_r^P) \tag{5}$$

Similarly

$$\sigma_0 = \alpha_1\varepsilon_0 + \alpha_2\varepsilon_r - \alpha_3(\varepsilon_0^P + d\varepsilon_0^P) \tag{6}$$

where

$$\alpha_1 = \frac{1 - v}{(1 + v)(1 - 2v)} \qquad \alpha_2 = \frac{v}{(1 + v)(1 - 2v)} \qquad \alpha_3 = \frac{1}{1 + v}$$

Substitution of eqns. (5) and (6) into (1) eliminates the stresses. Use of strain compatibility

$$\varepsilon_r = r\, d\varepsilon_0/dr + \varepsilon_0 \tag{7}$$

finally gives a second-order ordinary differential equation in ε_0:

$$r^2\alpha_1\, d^2\varepsilon_0/dr^2 + 3\alpha_1 r\, d\varepsilon_0/dr$$

$$= \alpha_3(\varepsilon_r^P + d\varepsilon_r^P) - \alpha_3(\varepsilon_0^P + d\varepsilon_0^P) + r\alpha_3 d/dr(\varepsilon_r^P + d\varepsilon_r^P)$$

or

$$r^2\, d^2\varepsilon_0/dr^2 + 3r\, d\varepsilon_0/dr = \delta\Sigma_r - \delta\Sigma_0 + d\Sigma_r/dr \tag{8}$$

where

$$\Sigma_r = (\varepsilon_r^P + d\varepsilon_r^P) \qquad \Sigma_0 = (\varepsilon_0^P + d\varepsilon_0^P) \qquad \delta = \frac{\alpha_3}{\alpha_1} = \frac{1 - 2v}{1 - v}$$

Equation (8) can be integrated twice to give an expression for ε_0:

$$r\, d\varepsilon_0/dr + 2\varepsilon_0 = \int_1^r \frac{\delta(\Sigma_r - \Sigma_0)}{r}\, dr + \delta\Sigma_r + C_1$$

$$\varepsilon_0 = \frac{1}{2}\int_1^r \frac{\delta(\Sigma_r - \Sigma_0)}{r}\, dr + \frac{1}{2r^2}\int_1^r r\delta(\Sigma_r + \Sigma_0)\, dr + C_1/2 + C_2/r^2$$

$$+ 1/r^2 \int_1^r r\delta\Sigma_r\, dr = u/r \tag{9}$$

Thus

$$u = r/2 \int_1^r \frac{\delta(\Sigma_r - \Sigma_0)}{r} \, dr + \frac{1}{2r} \int_1^r r\delta(\Sigma_r + \Sigma_0) \, dr + \frac{1}{r} \int_1^r r\delta\Sigma_r \, dr$$
$$+ \, C_1 r/2 + C_2/r \qquad (10)$$

From eqn. (7), using (9),

$$\varepsilon_r = \varepsilon_0 + 2\delta\Sigma_r - \frac{1}{r^2} \int_1^r \delta r(\Sigma_r + \Sigma_0) \, dr - 2/r^2 \int_1^r r\delta\Sigma_r \, dr - 2C_2/r^2 \qquad (11)$$

BOUNDARY CONDITIONS

Let the material parameters in the elastic inclusion be v and E'. The stresses in the inclusion are $\sigma_0 = \sigma_r$ throughout. The boundary conditions to be satisfied at the inclusion/matrix interface are that the radial stress and displacement should be continuous. Thus *in* the inclusion

$$\varepsilon_0 = E/E'[\sigma_0 - v(\sigma_r + \sigma_z)] \qquad (\varepsilon \equiv E\varepsilon/\sigma_Y)$$

But since $\varepsilon_z = 0$,

$$\sigma_z = v(\sigma_r + \sigma_0) \qquad \varepsilon_0 = E/E'\sigma_r(1 + v)(1 - 2v) \qquad \text{and} \qquad \sigma_r = \sigma_0$$

and at $r = 1$,

$$\varepsilon_{0_1} = u_1 = E/E'\sigma_{r_1}(1 + v)(1 - 2v) \qquad (12)$$

Now at $r = 1$, from (5)

$$\sigma_{r_1} = \alpha_1\varepsilon_{r_1} + \alpha_2\varepsilon_{0_1} - \alpha_3\Sigma_{r_1}$$

in the matrix

$$= \alpha_1(C_1/2 - C_2 + 2\delta\Sigma_{r_1}) + \alpha_2(C_1/2 + C_2) - \alpha_3\Sigma_{r_1} \qquad (13)$$

From (1) at $r = 1$

$$C_2 = -C_1/2 + u_1$$

which with (12) and (13) gives

$$C_2 = -C_1/2 + E/E'(1 + v)(1 - 2v)[(\alpha_1 + \alpha_2)C_1/2 + (\alpha_2 - \alpha_1)C_2$$
$$+ \, \Sigma_{r_1}(2\delta\alpha_1 - \alpha_3)]$$

or

$$C_2[1 - (2v - 1)E/E'] = C_1/2[E/E' - 1] + E/E'(1 - 2v)\Sigma_{r_1}$$

or

$$C_2 = C_1/2(\beta_1) + \Sigma_{r_1}\beta_2 \qquad (14)$$

where

$$\beta_1 = \frac{(E/E' - 1)}{[1 - (2v - 1)E/E']}$$

$$\beta_2 = \frac{(1 - 2v)E/E'}{[1 - (2v - 1)E/E']}$$

In order to separate the two constants C_1, C_2 the applied loading on the outer boundary must be considered. A finite and an infinite outer boundary were used and both are now considered.

(a) Finite Outer Boundary

Consider first the case where the applied stress is at $r = R$ and $\sigma_{r_R} = \sigma_0$. From (5):

$$\sigma_0 = \alpha_1 \varepsilon_{r_R} + \alpha_2 \varepsilon_{\theta_R} - \alpha_3 \Sigma_{r_R}$$

and using (9) and (11)

$$= \alpha_1 \left[\varepsilon_{\theta_R} + 2\delta\Sigma_{r_R} - 1/R^2 \int_1^R \delta_r(\Sigma_\theta + \Sigma_r)\,dr - \frac{2C_2}{R^2} - \frac{2}{R^2} \int_1^R r\delta\Sigma_r\,dr \right]$$

$$+ \alpha_2 \left[\tfrac{1}{2} \int_1^R \delta(\Sigma_r - \Sigma_\theta)/r \cdot dr + 1/2R^2 \int_1^R r\delta(\Sigma_r + \Sigma_\theta)\,dr \right.$$

$$\left. + 1/R^2 \int_1^R r\delta\Sigma_r\,dr + C_1/2 + C_2/R^2 \right] - \alpha_3 \Sigma_{r_R}$$

Using (14) to eliminate C_2 gives

$$\sigma_0 = (\alpha_1 + \alpha_2) \left[\delta/2 \int_1^R (\Sigma_r - \Sigma_\theta)/r\,dr + \delta/2R^2 \int_1^R r(\Sigma_r + \Sigma_\theta)\,dr \right.$$

$$\left. + \delta/R^2 \int_1^R r\Sigma_r\,dr + C_1/2 + \beta_1/R^2 \cdot C_1/2 + \beta_2/R^2\Sigma_{r_1} \right]$$

$$+ \alpha_1 \left[2\delta\Sigma_{r_R} - \delta/R^2 \int_1^R r(\Sigma_\theta + \Sigma_r)\,dr - 2\delta/R^2 \int_1^R r\Sigma_r\,dr \right.$$

$$\left. - 2\beta_1/R^2 \cdot C_1/2 - 2\beta_2/R^2\Sigma_{r_1} \right] - \alpha^3\Sigma_{r_R}$$

Rearranging terms gives

$$C_1/2[(\alpha_1 + \alpha_2) + (\beta_1/R^2)(\alpha_2 - \alpha_1)] = \sigma_0 - \Sigma_{r_R}(2\delta\alpha_1 - \alpha_3)$$

$$- \delta/2(\alpha_1 + \alpha_2) \int_1^R \frac{\Sigma_r - \Sigma_\theta}{r} \, dr - \frac{\delta}{2R^2} (\alpha_2 - \alpha_1) \int_1^R (3\Sigma_r + \Sigma_\theta) \, dr$$

$$- \beta_2/R^2 \Sigma_{r_1}(\alpha_2 - \alpha_1)$$

or

$$C_1 = \beta_R \left[2\sigma_0 - \delta(\alpha_1 + \alpha_2) \int_1^R \frac{\Sigma_r - \Sigma_\theta}{r} \, dr - \frac{\delta(\alpha_2 - \alpha_1)}{R^2} \int_1^R r(3\Sigma_r + \Sigma_\theta) \, dr \right.$$

$$\left. - 2\beta_2/R^2 \, \Sigma_{r_1}(\alpha_2 - \alpha_1) - 2\alpha_3\Sigma_{r_R} \right] \quad (15)$$

where

$$\beta_R = 1/[(\alpha_1 + \alpha_2) + (\beta_1/R^2)(\alpha_2 - \alpha_1)]$$

Elastic Solution
From (9)

$$\varepsilon_\theta = C_1/2 + C_2/r^2 \quad (16)$$

also

$$\varepsilon_r = C_1/2 - C_2/r^2$$

The constants C_1 and C_2 are found from (14) and (15):

$$C_1 = 2\beta_R\sigma_0 \quad \text{and} \quad C_2 = C_1/2\beta_1$$

Thus the stresses are found from (5) and (6)

$$\sigma_r = \alpha_1\varepsilon_r + \alpha_2\varepsilon_\theta$$

$$\sigma_\theta = \alpha_1\varepsilon_\theta + \alpha_2\varepsilon_r$$

$$\sigma_z = v(\sigma_r + \sigma_\theta) = \frac{C_1 v}{2}(\alpha_1 + \alpha_2)$$

$$= \text{constant} \quad (17)$$

Plastic Solution
For ease of numerical solution eqns. (9) and (11) can be rearranged to give

$$\varepsilon_\theta = \delta/2\Gamma_1 + \delta/2r^2\Gamma_2 + \delta/2D\Gamma_1 + \delta/2r^2D\Gamma_2 + C_1/2 + C_2/r^2$$

$$\varepsilon_r = \varepsilon_\theta + 2\delta\varepsilon_r^P + 2\delta \, d\varepsilon_r^P - \delta/r^2\Gamma_2 - \delta/r^2D\Gamma_2 - 2C_2/r^2$$

where

$$C_1 = \beta_R[2\sigma_0 - \delta(\alpha_1 + \alpha_2)I_1^R - \delta(\alpha_2 - \alpha_1)/R^2 I_2^R$$
$$- \delta(\alpha_1 + \alpha_2)DI_1^R - \delta(\alpha_2 - \alpha_1)/R^2 DI_2^R$$
$$- 2\beta_2/R^2(\alpha_2 - \alpha_1)(\varepsilon_{r_1}^P + d\varepsilon_{r_1}^P) - 2\alpha_3(\varepsilon_{r_R}^P + d\varepsilon_{r_R}^P)] \qquad (18)$$

and

$$C_2 = (C_1/2)\beta_1 + \beta_2(\varepsilon_{r_1}^P + d\varepsilon_{r_1}^P)$$

also

$$I_1^i = \int_1^{r_i} \frac{\varepsilon_r^P - \varepsilon_\theta^P}{r}\,dr \qquad I_2^i = \int_1^{r_i} r(3\varepsilon_r^P + \varepsilon_\theta^P)\,dr$$

$$DI_1^i = \int_1^{r_i} \frac{d\varepsilon_r^P - d\varepsilon_\theta^P}{r}\,d\theta \qquad dI_2^i = \int_1^{r_i} r(3d\varepsilon_r^P + d\varepsilon_\theta^P)\,dr$$

(b) Infinite Outer Boundary

Consider the case where the applied stress σ_0 is at $r = \infty$ where it is assumed to be always elastic. At $r = \infty$ from (5):

$$\sigma_0 = \alpha_1\varepsilon_{r_\chi} + \alpha_2\varepsilon_{\theta_\chi} - \alpha_3\Sigma_{r_\chi}$$

$$= \alpha_1(\varepsilon_0) + \alpha_2\left(1/2\int_1^\chi \delta(\Sigma_r - \Sigma_0)/r\,dr + C_1/2\right) - \alpha_3(0)$$

$$C_1 = 2\sigma_0/(\alpha_1 + \alpha_2) - \delta\int_1^\chi (\Sigma_r - \Sigma_0)/r\,dr$$

$$= 2\sigma_0/(\alpha_1 + \alpha_2) - \delta I_1^\chi - \delta DI_1^\chi \qquad (19)$$

The elastic and plastic solutions are found in an analogous way to the case of the finite outer boundary with the exception that eqn. (19) is used to determine the value of C_1.

Rigidity of Inclusion

The relative rigidity of the inclusion and the matrix are controlled by the E/E' ratio. For the rigid inclusion $E/E' = 0$ and thus $\beta_1 = -1$, $\beta_2 = 0$.

When E' drops to 0 then the inclusion becomes a hole, with $E/E' \to \infty$. Thus $\beta_1 \to 1/(1 - 2v)$, $\beta_2 \to +1$.

METHOD OF SOLUTION

The numerical solution of eqn. (9) requires the discretisation of the strain field, which in this case due to symmetry is one-dimensional and characterised by a radial line (Fig. 1). Two outer boundary configurations are used—finite and infinite—and in both cases 51 discrete points were taken. In the case of the finite boundary equal divisions were taken between points but in the infinite case, in order to cover as much of the field, the radii of the points were graded such that for the point $i + 1$

$$r_{i+1} = Tr_i \qquad \text{and} \qquad T = \sqrt{1 \cdot 125}$$

The outermost point in the infinite boundary case lies at radius $r/a = 18 \cdot 419$, where a is the radius of the inclusion. The actual procedure used is illustrated in the following:

1. An elastic solution is obtained with an applied stress $\sigma_0 = 1$ either (i) by using analytical expressions to determine stresses and strains explicitly or (ii) by determining the constants C_1, C_2 and then using eqns. (16) and (17). Stresses and strains are thus found at each point.

2. The elastic solution is scaled up to give initial yield. The applied loads $\sigma_0/\sigma_y|$ yield for each case are given in Table 1. The integrals of accumulated plastic strains I_1^i, I_2^i are zeroed.

3. The applied load is incremented and initial estimates of the new values of $\varepsilon_r, \varepsilon_\theta$ made at each point.

4. Using the Mendelson technique the current increments of component plastic strains ($d\varepsilon_\theta^P$, etc.) are computed from $\varepsilon_\theta, \varepsilon_r$.

5. The integrals DI_1, DI_2 are determined at each point by numerical integration.

6. The constants C_1, C_2 are found using eqn. (18) (finite boundary case) or eqn. (19) (infinite boundary case) and eqn. (14).

7. Re-estimates of ε_θ, ε_r at each point are calculated using eqns. (9) and (11).

8. Steps 4–8 are repeated 20 times to allow convergence.

9. Stresses $\sigma_r, \sigma_\theta, \sigma_z$ are calculated at each point using eqns. (5), (6) and (3).

10. Plastic strains are incremented ($\varepsilon_\theta^P + d\varepsilon_\theta^P$) and the integrals I_1^i, I_2^i determined at each point, in preparation for the next load increment.

11. Steps 3–11 are repeated until the required number of load increments are completed.

TABLE 1

COMPARISON OF $\gamma_{max}/\gamma_{max}\,R|_{interface}$ WITH Q AT 3200% LOAD

Outer radius	E/E'	$\sigma_0/\sigma_Y|_{yield}$	Elastic	3200%	
				$Q = 0.0$	$Q = 0.025$
∞	0	1·250	2·000	5·344	5·082
	0·5	2·027	1·417	2·753	2·679
	1·0	2·500	1·000	1·487	1·468
	10·0	0·764	2·800	2·125	1·795
6·0	0	1·264	1·946	4·34	4·15
	0·5	2·037	1·400	2·237	2·188
	1·0	2·500	1·000	1·21	1·19
	10·0	0·748	2·667	1·767	1·807

A total of 20 load increments was required to reach a final load of 32 times initial yield load which is denoted 100%. The first four were incremented in steps of 50%, followed by three at 100% and finally 13 at 200%. The numerical integration employed was the simple trapezoidal rule.

RESULTS

The purpose of the present work was (a) to determine the rate of increase of stress and strain at the inclusion interface with increasing applied stress, and (b) determine stress and strain concentration factors.

The analysis was formulated such that the outer boundary may be taken as infinite and elastic or finite and elasto-plastic. Solutions were obtained for finite outer boundaries of $R/a = 3·0, 6·0, 11·0$ for a rigid inclusion and a work-hardening index of $Q = 0·0$. The stress concentration factors at the maximum load were within $\pm 3\%$ of each other and so an outer radius of 6·0 was used thereafter as representative of the finite boundary case.

The applied load σ_0/σ_Y, to cause initial yield with a Poisson's ratio of 0·3 for each case of outer boundary and inclusion rigidity, is shown in Table 1, and in each case this load is indicated as 100%. The load is then incremented in 20 graded steps up to 3200%. The graphs in Figs. 3–6 are drawn against EPO, where

$$EPO = E \int d\bar{\varepsilon}^P / \sigma_Y|_{outer\ boundary}$$

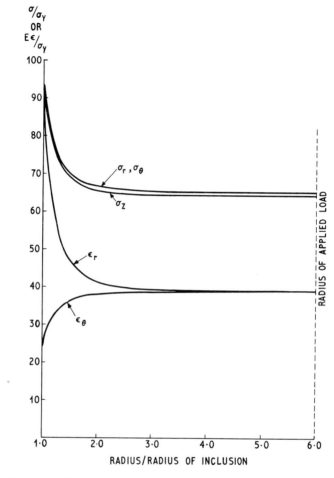

FIG. 2.   Example of stress and strain distributions $E/E' = 0.5$, $Q = 0$, $R/a = 6.0$,
3200% load.

The outer boundary is at $r/a = 6.0$ for the finite case, and for the infinite case is taken as radius $r/a = 18.25$, which is the outermost point of the grid.

The maximum stress at the interface is the radial stress σ_r for cases $E/E' = 0.0, 0.5, 1.0$, but for the 'soft' inclusion ($E/E' = 10.0$), σ_θ is greater than σ_r. Similarly for the harder inclusion cases, the maximum normal strain both at the surface and the outer boundary is ε_r and so the maximum shear strain is given by ε_r since $\varepsilon_z = 0$. In the case of the soft inclusion, the

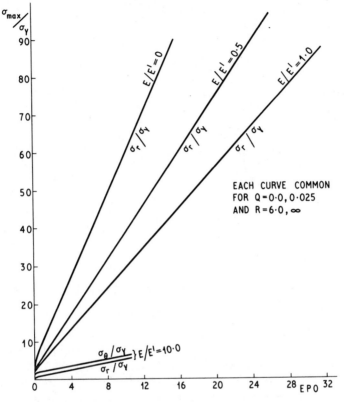

FIG. 3. Development of maximum interfacial stress.

maximum strain changes to ε_θ and so γ_{max} is given by ε_θ. When the infinite outer boundary is considered, the two normal strains are equal and so

$$\left. \frac{E\gamma_{max}}{\bar{\sigma}_Y} \right|_\infty = 0.52\sigma_0/\sigma_Y$$

Figure 2 is given as an example of the stress and strain distributions for an elastic inclusion $(E/E' = 0.5)$ in a matrix with $Q = 0.0$ under a load of 3200% applied at radius 6.0. It is clear that the effect of the inclusion on the stress and strain fields is restricted to about two inclusion radii from the interface.

Development of Stress and Strain with EPO

Figures 3 and 4 show consistent build-up of stress and strain with applied

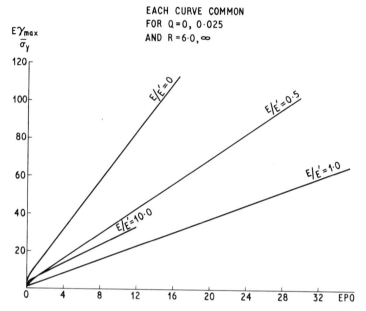

FIG. 4. Development of maximum interfacial shear strain.

load. For each curve of E/E' the four curves for $Q = 0\cdot0$, $0\cdot025$ and outer radius 6, ∞ are almost identical and only one curve is drawn. The unusual curve for $E/E' = 10\cdot0$ (Fig. 4) is due to the maximum interfacial normal strain changing from ε_r to ε_θ. The effect of the soft inclusion is also to make $\sigma_\theta > \sigma_r$ and so both curves are shown on Fig. 3.

Development of Stress and Strain Concentration Factors with EPO

In contrast to the shear strain and maximum stress at the interface, Figs. 5 and 6 show that the stress and strain concentration factors reach near limiting values by $EPO = 15$. The curves are drawn for $Q = 0$ and comparison with values for $Q = 0\cdot025$ is shown in Tables 1 and 2 at a load of 3200%. The curves for $Q = 0\cdot025$ also show the same limiting trend. The reversal of limiting trend for $E/E' = 10\cdot0$ is as explained in the last section.

Comparisons with Orr and Brown

Figures 3–6 show similar trends to Orr and Brown of the steady rise with EPO of maximum interfacial stress and of maximum interfacial shear strain, a common curve being exhibited for both $Q = 0$ and $0\cdot025$. In Orr

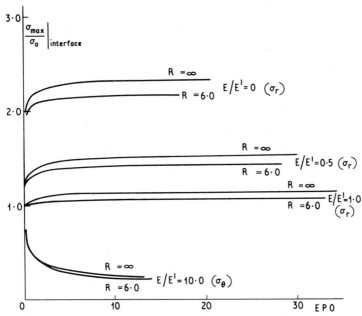

FIG. 5.   Development of stress concentration factor for $Q = 0.0$.

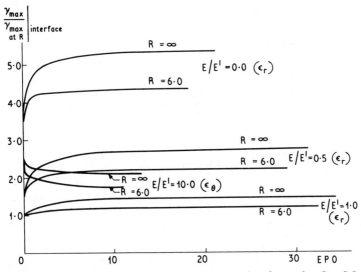

FIG. 6.   Development of shear strain concentration factor for $Q = 0.0$.

TABLE 2

COMPARISON OF $\sigma_{max}/\sigma_0|_{interface}$ WITH Q AT 3200 % LOAD

Outer radius	E/E'	$\sigma_0/\sigma_y|_{yield}$	Elastic	3200 %	
				$Q = 0.0$	$Q = 0.025$
∞	0	1·250	1·4	2·33	2·25
↓	0·5	2·027	1·167	1·535	1·511
	1·0	2·500	1·0	1·148	1·141
↓	10·0	0·764	1·720	0·260	0·294
6·0	0	1·264	1·38	2·17	2·10
↓	0·5	2·037	1·161	1·434	1·413
	1·0	2·500	1·0	1·073	1·069
↓	10·0	0·748	1·755	0·249	0·283

and Brown, the technique of solution was to initially assume that the matrix and the inclusion had $E/E' = 1.0$ up to overall yield, after which the inclusion was held rigid. From the trend of stress and shear strain concentration ratios quoted therein and using Figs. 5 and 6, it would appear that the effective rigidity of the inclusion was of the order $E/E' = 0.5$.

CONCLUSIONS

As predicted by Orr and Brown, the interfacial stresses and shear strains continue to rise with applied stress on the distant boundaries even for $Q = 0.0$. The stress and shear strain concentration ratios indicate a limiting behaviour, the limits being more dependent on inclusion rigidity than radius of outer boundary or matrix work-hardening.

REFERENCES

1. ORR, J. and BROWN, D. K. Elasto-plastic solution for a cylindrical inclusion in plane strain, *Engng Frac. Mech.*, **6** (1974), pp. 261–74.
2. ARGON, A. S., IM, J. and SAFOGLU, R. Cavity formation from inclusions in ductile fracture, *Metallurgical Trans.*, **6** (1975), pp. 825–37.
3. DAVIS, E. A. Extension of iterative method for determining strain distributions to the uniformly stressed plate with a hole, *J. Appl. Mech.*, **30** (1963), No. 2, pp. 210–14.

4. ROBERTS, E. and MENDELSON, A. Analysis of Plastic Thermal Stresses and Strains in Finite Thin Plate of Strain Hardening Material, NASA TN D-2206 (1964).
5. MENDELSON, A. *Plasticity: Theory and Applications*, Macmillan, London, 1968.
6. MENDELSON, A. Elastoplastic Analysis of Circular Cylindrical Inclusion in Uniformly Stressed Infinite Homogeneous Matrix. NASA TN-D4350 (1968).

Chapter 11

FORMATION OF PERMANENTLY CURVED BORON FILAMENTS

M. Schoppee & J. Skelton

FRL, Massachusetts, USA

SUMMARY

Permanently curved configurations are produced when bent boron filaments are exposed to high temperatures. Filament modulus is maintained, but some strength loss is observed after exposure to temperatures above 600°C. A radius of curvature as low as 0·7 cm can be set in 100 µm diameter boron filaments with a strength penalty of 30%.

INTRODUCTION

Boron is unusual among high temperature, high performance fibres because it deforms plastically at elevated temperatures considerably below its melting temperature. Consequently, bending and tensile stresses can be relieved to a significant degree by application of heat to this material without markedly altering its characteristics of high strength and high modulus, thereby making possible the production of relatively stress-free boron-filament-reinforced composite structures in a curved form. Individual curved boron filaments could find application as needles for microsurgery where maximum rigidity and strength are required.[1] During the course of a feasibility study of this proposed application, techniques were developed and data were obtained relating to the settability of large bending deformations in boron filaments over a range of elevated temperatures. The effects of heat treatments in air on the filament strength and modulus at ambient temperature after exposure and the ultimate strains sustainable by bent filaments at various temperatures were also explored.

227

The phenomenon of significant heat relaxation in boron filaments was reported by Prewo,[2] who observed relief of bending strains as high at 70 % of the applied strain at 850 °C in vacuum, a strain relief which can only result from plastic deformation. The linear tensile stress–strain behaviour of boron fibres at ambient temperature, however, provides no evidence of plasticity[3] and boron filaments exhibit only very low levels of primary creep at ambient temperature. Ericksen[4] observed a logarithmic creep coefficient of 0·001–0·002 %/h for applied strains in the range 0·2–0·5 %. There is, however, considerable evidence of plasticity at elevated temperatures: Ellison and Boone[5] measured both tensile and creep properties of boron filaments at high temperatures in vacuum and found plastic tensile deformations greater than 0·5 % at temperatures between 650 °C and 815 °C and primary creep of 0·7 % within a few minutes after application of tensile strains of 0·5 % at 815 °C; high levels of primary creep were also reported by Antony and Chang,[6] who measured 0·1–0·15 % primary creep at 260 °C under an applied strain of 0·2 % and by Rose and Stokes,[7] who observed 0·2 % primary creep under an applied strain of 0·8 % at 540 °C.

This current paper is concerned with the establishment of the degree of curvature which can be achieved by heat relaxation in a single boron filament and the concomitant reduction in strength and modulus of the filament over a range of exposure conditions. This information is needed to establish optimum processing conditions for producing permanently curved boron fibres for particular end-use applications.

EXPERIMENTAL PROCEDURE AND RESULTS

Boron Filaments

The materials used in this study were 100 μm diameter boron filaments manufactured by Composite Materials Corp.; some properties of these fibres at ambient temperature are listed in Table 1. The elastic modulus is quoted from the manufacturer's data; the rupture stress value was determined by tensile testing 2·5 cm specimens bonded to stiff tabs with epoxy cement; the rupture strain was then calculated from the stated modulus and measured rupture stress assuming a linear relationship between stress and strain to failure. While the straightforward tensile test technique used may not yield as high values of rupture stress as more sophisticated techniques involving sequential testing of increasingly shorter lengths of the same fibre coupled with subsequent statistical analysis of results,[8] it is, nevertheless, adequate for determinations of relative strength

TABLE 1
PROPERTIES OF BORON FILAMENTS AT $20\,°C$

Diameter	$100\,\mu m$ ($0\cdot0040$ in)
Specific gravity	$2\cdot6$
Elastic modulus	$41 \times 10^{10}\,\mathrm{Nm}^{-2}$ (58×10^6 psi)
Tensile rupture stress	$3\cdot9 \times 10^9\,\mathrm{Nm}^{-2}$ ($5\cdot6 \times 10^5$ psi)
Tensile rupture strain	$1\cdot0\%$
Limiting loop radius of curvature	$0\cdot51$–$0\cdot36$ cm ($0\cdot20$–$0\cdot14$ in), average $0\cdot42$ cm ($0\cdot17$ in)
Limiting bending strain	$1\cdot0$–$1\cdot4\%$, average $1\cdot2\%$

loss and modulus change after exposure to high temperature. Similarly, although scatter in test results involving boron fibre is often quite large, the data reported herein will, in general, represent average values of multiple tests; no attempt has been made to deal with the statistical implications of the data.

Applied Bending Strains
Bending strains were induced in the boron filaments by tying in them a simple overhand knot. The loop formed in this manner assumes a nearly circular geometry, and, consequently, the bending strain imposed on the fibre by the bent configuration is uniform over the length of fibre in the loop; the maximum imposed bending strain ε_B occurs at the fibre surfaces on the inside and outside of the bend and is given by the expression

$$\varepsilon_B = r/R_0 \tag{1}$$

where r is the radius of the fibre and R_0 is the radius of curvature of the loop. The term bending strain will be used throughout this paper to mean maximum bending strain occurring at the fibre surface. Derivation of the above expression, and those to follow which are similar to it, assume that the longitudinal stress–strain properties of the fibre in compression are symmetrical with respect to those in tension, so that no translation of the neutral axis from the axis of symmetry occurs during bending.[9] The *imposed* surface bending strain given by eqn. (1) is independent of and in addition to any residual longitudinal strains which may have arisen in the fibre during manufacture,[10] since if these residual strains exist they must necessarily be symmetrically distributed about the fibre axis.

The limiting loop radius of curvature of the boron filaments reported in Table 1 was determined by pulling knotted loops progressively tighter until rupture occurred and measuring the loop radius at rupture. The limiting bending strains associated with these values of loop radius of curvature by eqn. (1) are also noted in Table 1. The gauge length in this type of rupture strain determination is the loop circumference which in this case varies between 2·3 cm and 3·2 cm—approximately the same magnitude as the gauge length employed in the tensile testing. The tensile and bending rupture strain values are thus directly comparable and, while the values overlap to some extent, the bending strain values are generally higher probably because maximum strain is imposed on a lesser portion of the fibre during bending than during axial tensioning.

Heat Source

The high temperatures necessary to achieve heat relaxation in bent boron filaments were attained by exposing them to a bilateral radiant heat flux supplied by two opposing quartz-faced radiant heater panels. These panels measure 30 cm by 15 cm and were spaced 1·25 cm apart; the temperature field over the central 20 cm by 10 cm area between these heaters is quite uniform. It is estimated that exposed filamentary specimens reach the equilibrium temperature of the heater surfaces within 30 s of the onset of exposure. Temperatures as high as 1300 °C can be achieved with this heater configuration.

Heat-Setting of Bent Filaments

Settability of the bent form in boron exposed at various temperatures was determined by positioning knotted, looped filaments of various radii of curvature between the heater faces for various lengths of time, allowing them to cool after withdrawal from the heat, releasing the knot, and measuring the radius of curvature of the recovered bent filament. The amount of imposed bending strain relieved by the application of heat is equivalent to that which is not recovered upon release of the knot; the unrecovered bending strain ε_u is related to the radius of curvature of the bent filament after exposure R_1 by the following expression similar to eqn. (1):

$$\varepsilon_u = \frac{r}{R_1} \tag{2}$$

The degree of set $S(T)$ of the imposed bending deformation at a particular setting temperature T is equivalent to the fractional unrecovered fibre bending strain, that is, the ratio of unrecovered bending strain ε_u to applied bending strain ε_B,

$$S(T) = \frac{\varepsilon_u}{\varepsilon_B} = \frac{r/R_1}{r/R_0} = R_0/R_1 \tag{3}$$

The effect of duration of exposure at two selected temperatures on fractional unrecovered bending strain is illustrated in Fig. 1 for an applied strain of 0·4 %. For these determinations fibre loops of 1·25 cm radius were exposed at temperature for the indicated time, allowed to cool for 1 min prior to releasing the knot forming the loop, and then allowed to recover for 1 min before the recovered radius of curvature was measured. As shown in

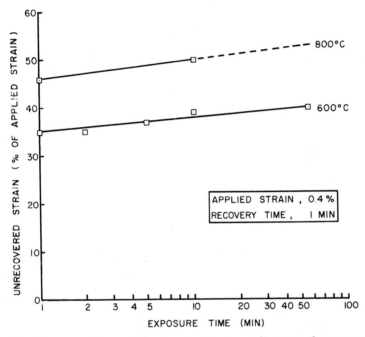

FIG. 1.   Unrecovered imposed bending strain in boron filaments after exposure at elevated temperature for various lengths of time.

Fig. 1, the greatest portion of strain relief occurs during the first minute of exposure; subsequent relief occurs at the rate of 3–4 % of the applied strain per decade. Recovery is even less affected by extended recovery time periods: measurements of recovered radius of curvature on some specimens allowed to recover for 2 h showed no additional recovery after the first minute. Consequently, the following time standards were established for all subsequent testing: exposure time, 1 min; cooling time, 1 min; recovery time, 1 min.

The magnitude of the bending deformations which can be heat-set in

boron fibres over the temperature range 400–1100 °C according to the time
schedule outlined above, is shown in Fig. 2; the set, or unrecovered strain
expressed as a fraction of applied strain, rises nearly linearly with increasing
exposure temperature from 20 % at 400 °C to 80 % at 1100 °C. The effect of
level of applied bending strain on the degree of set was determined by heat
setting fibre loops of several different radii of curvature. As can be seen from

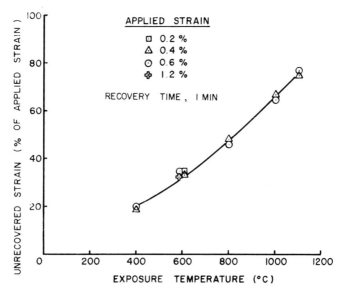

FIG. 2.   Unrecovered imposed bending strain in boron filaments after 1 min
exposure at elevated temperatures.

the data plotted in Fig. 2, the fractional unrecovered strain is independent
of the level of applied strain.

The levels of unrecovered strain shown in Fig. 2 are clearly sufficient to
produce boron filaments in a curved form; however, most practical
applications require that the strength and modulus values of the curved
fibres remain high, and these properties were explored in some detail.

Strength Loss in Heat-Exposed Fibres
Strength loss and modulus change in both straight and bent filaments
exposed to elevated temperatures from 600 °C to 1000 °C were measured in
order to ascertain not only the effect of heat alone but also the effect of heat
applied to filaments under strain. The same techniques were used to tensile

test the straight filaments after exposure as were used to determine the filament rupture stress in the as-received condition. The specimen test length of 2·5 cm employed in these tests is not sufficiently long to allow accurate direct determination of filament modulus; however, the slopes of the stress–strain curves for fibres exposed at different temperatures may be compared to obtain relative moduli. The moduli of straight specimens

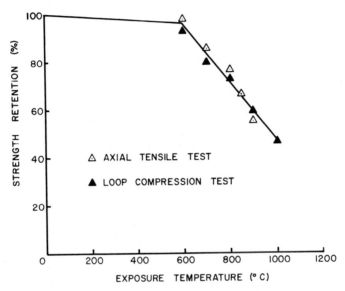

FIG. 3.   Strength retention in boron filaments at ambient temperature after exposure at elevated temperature in air for 1 min.

exposed to temperatures from 600 °C to 900 °C in air for 1 min and then cooled to ambient temperature were the same as the modulus of the unexposed filament. The rupture strength of these straight filaments tested in axial tension decreased linearly with increasing exposure temperatures above 600 °C in the manner shown in Fig. 3; the rupture strength falls to approximately 60 % of its original value after exposure at 900 °C.

The relative rupture stress and modulus of knotted loops of fibre exposed over the same temperature range were determined by means of a loop compression test. Exposed specimen loops of 0·6 cm radius of curvature (0·8 % applied bending strain), in which the constraining knots were not released to allow recovery, were bonded to stiff tabs as illustrated in Fig. 4. Epoxy cement was applied to the knotted section of each filament in a

manner which ensured that the same length of loop arc was encompassed in
the bonded area for each test specimen. These filament loops were
compressed to rupture against a flat plate in a tensile test machine and
stress–deflection curves were generated. Differences in slope of these curves
could only result from differences in filament modulus since the geometry of
each test specimen was the same. As in the axial tensile tests, no change in
filament modulus was observed as the result of exposure to temperatures
over the range 600–1000 °C.

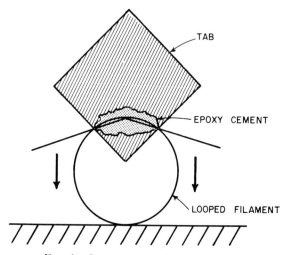

FIG. 4.   Loop compression test.

The axial fibre tensile stress at which the compressed fibre loops ruptured
can be estimated by considering that the test configuration approximates
that of an elastica loop compressed between two parallel plates. In this case,
the maximum radius of curvature ρ_2 of the compressed elastica is related to
the plate separation D by the expression[11]

$$\rho_2 = D/2\cdot4 \tag{4}$$

Hence, the maximum strain ε_m imposed on the bent elastica is given by the
relationship

$$\varepsilon_m = \frac{r}{\rho_2} = \frac{2\cdot4r}{D} \tag{5}$$

In the loop compression test the imposed surface strain at rupture for
unheat-set fibre loops may be approximated by this same expression, where

D represents the loop height at rupture. In the exposed fibre loops, however, that increment of the applied strain which has been relieved by heat, ε_u, must be subtracted from the strain given by eqn. (5) so that the imposed surface strain at rupture for these filaments is given by

$$\varepsilon_m = \frac{2 \cdot 4r}{D} - \varepsilon_u = \frac{2 \cdot 4r}{D} - \frac{rS(T)}{R_0} \tag{6}$$

from eqns. (2) and (3). Since no change in filament modulus was observed as the result of exposure at elevated temperatures, the relative strengths of the filaments heated in the bent form is given by the ratio of imposed strain at rupture for the heat-set fibre loop, eqn. (6), to the imposed strain at rupture of the unheat-set fibre loop, eqn. (5).

The relative strengths of boron filaments in which bending deformations have been heat-set as determined by the loop compression test in the above manner are also plotted in Fig. 3; they are seen to be in excellent agreement with those values of relative strength determined for straight filaments exposed to the same heat conditions. Therefore, the strength retention of heat-set fibres is affected only by the conditions of exposure at elevated temperature and not by the state of strain of the fibre during exposure.

Minimum Radius of Curvature

The limiting radius of curvature to which the boron filaments may be bent at ambient temperature is defined by the imposed bending strain at rupture of the filaments by eqn. (1). The average limiting radius of curvature given in Table 1 is $0 \cdot 42$ cm, corresponding to an average imposed bending strain at rupture of $1 \cdot 2 \%$. However, fibre loops of this radius cannot be heat-set at temperatures above 800 °C because the loop ruptures spontaneously within a few seconds of initiation of exposure at these temperatures. Such spontaneous failure results from the inability of the heating fibre to sustain the residual strain in the filament as it increases in temperature. The residual strain ε_R in a bent fibre at any temperature T is related to the applied strain ε_B and the strain relieved at that temperature ε_u by the following expression derived from eqns. (1), (2) and (3):

$$\varepsilon_R = \frac{r}{R_0} - \frac{r}{R_1} = -\frac{r}{R_0}[1 - S(T)] \tag{7}$$

The level of residual bending strain in bent boron filaments as it is affected by a 1 min exposure at various temperatures in air is illustrated in Fig. 5 for a range of applied strains; the pertinent values of $S(T)$ necessary to the calculation were taken from Fig. 2.

The maximum sustainable level of residual imposed bending strain at any particular temperature may be determined by exposing fibre loops of several radii of curvature at that temperature and noting the radius of the smallest loop which survives a 1 min exposure. The difficulty in this approach, however, is that the smallest loops which can be formed at ambient temperature are not sufficiently small to fail spontaneously at

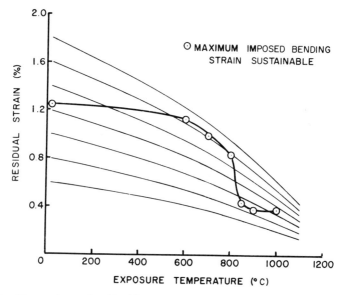

FIG. 5.   Comparison of residual imposed bending strain with maximum imposed bending strain sustainable by boron filaments at elevated temperatures.

temperatures below 800 °C. Consequently, smaller loops were formed sequentially at elevated temperatures below the desired heat-setting temperature. For instance, loops with a radius as small as 0·31 cm can be formed at 600 °C by drawing an initially larger loop increasingly tighter during exposure. The fractional recovery of loops formed at elevated temperature and then heat-set was the same as that for loops formed at ambient temperature and then heat-set with no further drawing. Many of these small, heat-formed loops were subsequently heat-set at temperatures higher than the forming temperature to determine the average minimum radius which could survive for a range of setting temperatures between 600 °C and 1000 °C. The values of maximum sustainable imposed bending strain associated with these minimum radii of curvature are superimposed on the curves of residual imposed strain in Fig. 5; they serve to define the lower limits of recovered radius of curvature which can be achieved by heat-setting of bent boron filaments.

These limits of recovered radius of curvature are shown in Fig. 6 as a function of heat-setting temperature; also shown is the corresponding loop radius which must be set to produce them. Since these plotted values

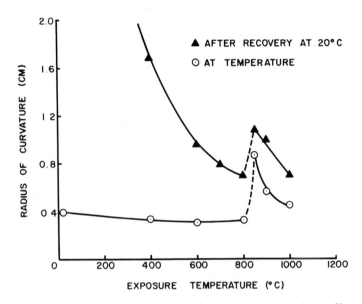

FIG. 6.   Minimum radius of curvature settable in 100 μm diameter boron filaments at various setting temperatures.

represent average behaviour, some fibres may be formed to slightly smaller radii of curvature than specified in Fig. 6, and some may fail during the heat-setting process. On the average, however, a permanently curved configuration of a radius of curvature of 0·70 cm can be achieved in a boron filament of the type used in this investigation by sequential formation of a fibre loop of radius 0·31 cm at temperatures below 800 °C followed by heat-relaxation at 800 °C. The same stable radius of curvature can also be achieved by heat-setting at 1000 °C a loop of 0·46 cm radius formed at ambient temperature.

Each of these processes involves a certain loss in strength of the boron filament but no decrease in modulus. Figure 7, a composite depiction of the information contained in Figs. 3 and 6, illustrates the magnitude of this

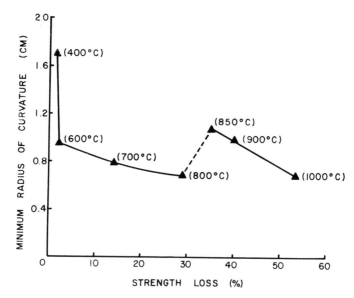

FIG. 7. Minimum radius of curvature achievable in boron filaments and associated strength loss.

strength loss over the range of permanent radii of curvature achievable. Curved filaments with a radius of 0·95 cm may be produced for virtually no strength loss penalty; fibres of a radius of curvature of 0·7 cm incur a strength loss penalty of 30% if set at 800 °C and 55% if set at 1000 °C. Presumably, curved fibres of even smaller radius of curvature can be produced by heat-setting at temperatures higher than 1000 °C but the cost in strength loss would be prohibitively large.

SUMMARY

Boron filaments can be heat-set to achieve a bent form over a range of temperatures considerably below the melting point. Bending deformations were set in looped specimens heated by exposure in air to a bilateral radiant heat source at temperatures ranging from 400 °C to 1100 °C. The degree of heat-relaxation of bending strains in 100 μm diameter fibres increases nearly linearly from 20% of the applied strain at 400 °C to 80% of the applied strain at 1100 °C after a 1 min exposure and 1 min recovery. Exposure times longer than 1 min produced only small increases in the amount of set; recovery times to 2 h result in no discernible increase in measured recovery. The degree of set, or fractional strain relief, is independent of the level of applied strain over the range 0·2–1·2 %.

Strength and modulus determination on both straight and bent filaments after heat exposure showed no change in modulus and a strength loss varying from only a few per cent after exposure at 600 °C to 55 % after exposure at 1100 °C. No effect on strength loss of the state of strain in the fibre at the time of exposure was observed.

Permanently curved forms of a radius of curvature as low as 0·7 cm can be produced in the boron filament by sequential heat forming of a fibre loop to 0·31 cm radius and subsequent heat-relaxation at 800 °C; the penalty in strength loss for achieving this configuration is 30 %. Curved boron filaments of a radius of curvature of 1·0 cm can be produced by heat-setting at 600 °C with virtually no loss in strength.

REFERENCES

1. ROBERTS, W. N. Development of needles and sutures for microsurgery, *J. Biomed. Mater. Res.*, **9** (1975), pp. 399–405.
2. PREWO, K. M. An elastic creep of boron fibers, *J. Composite Materials*, **8** (1974), pp. 411–14.
3. HERRING, H. W. Selected Mechanical and Physical Properties of Boron Filaments, NASA TN D-3202 (January 1966).
4. ERICKSEN, R. H. Room temperature creep and failure of boric filaments, *Fibre Sci. Technol.*, **7** (1974), pp. 175–83.
5. ELLISON, E. G. and BOONE, D. H. Some mechanical properties of boron–tungsten boride filaments, *J. Less Common Metals*, **13** (1967), pp. 103–11.
6. ANTONY, K. C. and CHANG, W. H. Mechanical Properties of Al–B Composites, Trans. ASM (1968), Vol. 61, pp. 550–8.
7. ROSE, F. K. and STOKES, J. L. Advanced Methods to Test Thin Gage Materials, AFML-TR-68-64 (April 1968).
8. PHOENIX, S. L. Statistical Analysis of Flaw Strength Spectra of High-Modulus Fiber, ASTM-STP 580 (1975).
9. SCHOPPEE, M. M. and SKELTON, J. Bending limits of some high-modulus fibres, *Textile Res. J.*, **44** (1974), pp. 968–75.
10. BEHRENDT, D. R. Longitudinal Residual Stresses in Boron Fibers, NASA-TM-X-73402 (1976).
11. SKELTON, J. The measurement of the bending elastic recovery of filaments, *J. Textile Inst.*, **56** (1965), T443–T453.

INDEX

Anisotropy of statically loaded
   composite plates, 55
Axial orientation, 182, 183
Azimuthal angle, 201, 202

Bending, rupture during, 17–35
Binary systems, 87, 107
Boron filaments
   bending strains in, 229
   heat relaxation in, 228, 230
   heat-setting, 230
   loop compression test, 233–4
   maximum imposed bending strain
     sustainable, 236
   minimum radius of curvature, 235,
     237, 238
   permanently curved configurations,
     227–9
   properties of, 228
   relative rupture stress, 233
   residual bending strain in, 235
   strength loss in heat-exposed fibres,
     232
   strength retention after exposure at
     elevated temperature, 233
   unrecovered imposed bending strain
     in, 231–2
Boundary value problem, 163
Bounding, 129

Bounding—contd.
   formulation, 151–9
     models, 159–62
     preliminary definitions, 151
   methods, 120, 121
   theory, 121
Bounds
   procedures of finding, 122
   technique of finding improved, 123
Bulk modulus, 166, 202, 204

Carbon fibre-reinforced phenolic
   resins, corrosion, 1–15
Carbon fibre-reinforced plastics, 2
Cartesian transformation, 181
Commutation between differential and
   averaging operators, 175
Compliance tensor, 184
Composite plates, effect of shear
   deformation and anisotropy of
   non-linear response of, 55–65
Constitutive relationships for
   heterogeneous materials,
   119–87
Constraint free formulation, 137
Continuous matrix, 165, 167
Correlation
   function, 153, 157, 159, 175
   symmetry, 163
   weighting function, 176

Corrosion
  assessment of, 3
  carbon fibre-reinforced phenolic
    resins, in, 1–15
Cylindrical shells, vibrations of
  antisymmetrically laminated,
  37–54

Dirac delta function, 138
Divergence theorem, 128, 136, 138,
  141

Effective elasticity, 154, 162
Effective moduli, 120, 163
Effective plastic strain remote from
  inclusion (EPO)
  development of stress and strain
    concentration factors with, 222
  development of stress and strain
    with, 221
Elasticity tensor, 184
Ellipsoidal symmetry, 149, 153, 157,
  163, 168, 177
Energy of extraction without shearing,
  28
Equilibrium strain energy, 128–31,
  134, 139, 141, 144, 154, 156, 161

Ferric chloride solution, 11
Fibre composites, strain
  measurements on, 190
Fibre layout, 22
Fibre–matrix interaction, 21
Fibre-reinforced composites, rupture
  during bending, 17–35
Fibre-reinforced plastics, 2
Fibre surface treatment effects, 13
Filament stiffened sheet, stress
  concentrations around circular
  holes in, 189–95
Force–deflexion curves, 19
Fourier
  space, 140, 142, 145, 147, 149, 150
  transforms, 142, 143, 145, 147–50,
    176

Friedel–Crafts resin, 5

Glacial acetic acid, 7
Green's tensor function, 137–42, 145,
  146, 176

Hermans orientation function, 183
Heterogeneous materials, constitutive
  relationships for, 119–87
Heterogeneous systems of statistical
  isotropy, 124
  variational formulation for, 124–6
  variational principles for, 172
Holes, circular, in filament stiffened
  sheet, stress concentrations
  around, 189–95
Hooke's law, 89

Inclusion/matrix interface, boundary
  conditions, 214
Inclusions
  elasto-plastic matrix, in, 209–25
  ellipsoidal, 185
  matrix structure, and, 165, 167
  rigidity of, 217
  shape role, 162–3, 165
  stress components in and around,
    198
Isotropic reference media, 146, 149

Jacobian, 150

Kronecker delta, 127

Lagrangian multipliers, 127, 134
Lamé constant, 146, 199
Lamination asymmetries, 37
Legendre polynomial, 199
Light field isochromatic patterns, 77
Linear Programming
  algorithm, 116
  problems, 104, 105, 114

Macroscopic scale, 151
Mendelson technique, 218
Micromechanics, 164
Microscopic scale, 151
Microstructure, 120, 122, 152, 168, 169, 170
Minimum strain criterion, 90
Model constitutive relationships, 164–8
Modelling estimates, 121
Models, bounding formulae as, 159–62
Multicomponent composites
  algebraic analysis, 100
  concentration extremes for conditions on reduced properties, 111
  constitutive relationships for estimating the properties of, 87
  constrained optimisation problem, 103
  constraining relationship for simple case, 105
  density and cost per pound, 90
  extension of binary constitutive relationships for transverse properties, 92
  extreme limits on concentration of components of ternary composite which satisfy property constraints, 113
  feasibility of current materials selection, 115
  input requirements, 114
  manifold of properties, 100
  optimum materials design for property (or characteristic), 115
  performance optimisation, 87, 96–100
  property optimisation analysis, 85–117
  reduced properties, 101
  specified property levels, 92–3, 97
  superimposed property maps, 97–8
  weight fraction of components to achieve specific property $\bar{P}$ at minimum cost, 111

Orientation
  averaging, 180
  axial, 182, 183
  invariants, 183
  random, 184
  relationships, 180
Orthotropic composites
  basic equations of elasticity, 69
  examples, 73
  numerical procedure, 70
  stress analysis, 67–83
Orthotropic half-plane subjected to concentrated load, 73
Orthotropic square plate model, 74

Parseval's relations, 147, 150
Performance optimisation, 86
  multicomponent composites, 87, 96–100
Phenol-formaldehyde novolak, 5
Phenolic resins, corrosion in carbon fibre-reinforced, 1–15
Photo-orthotropic elasticity, 67, 72
Piecewise constant, 159, 161
Plane strain problem, 209–25
Plastics, fibre-reinforced, 2
Plates, composite, effect of shear deformation and anisotropy on non-linear response of, 55–65
Poisson's ratio, 198, 201, 203, 206, 210, 212, 219
Preferred orientation, 124
Principle of
  complementary energy for systems subject to prescribed surface displacement, 129
  minimum potential energy, 127, 172
Property optimisation analysis for multicomponent (hybrid) composites, 85–117

Random orientation, 184
Reciprocal problem, 136
Reciprocal transformation, 135
Reciprocal variational problem, 129

Reference elasticity, 146, 155, 161, 162, 164–8
Reference systems, 140, 145
Reuss
    average, 156, 157, 159, 162, 184
    model, 160, 162, 168
Rule of mixtures, 87, 89
Rupture during bending, 17–35

Self-consistent method, 164–5
Shear
    deformation of statically loaded composite plates, 55
    modulus, 146, 166
    properties, limiting, 197
Shearing
    deformation of fibres, 22
    energy, calculation of, 33
Sheet, filament stiffened, stress concentrations around circular holes in, 189–95
Shells, cylindrical, vibrations of antisymmetrically laminated, 37–54
Shock fracture during bending, 17
Sodium hydroxide, 5–7
Space scales, 151
Statistical isotropy, 154
Statistical symmetries, 152, 153, 155, 157, 160, 170
Statistical symmetry descriptors, 162, 165
Steam, corrosion in, 12
Strain, 164
    compatibility parameter, 90
    concentration factor, 222
    distributions, 221
    energy, 161, 162
    field, 132–4, 145, 218
    incompatible failure criterion, 116
    measurements on fibre composites, 190
Stress
    analysis of orthotropic composites, 67–83
    components in and around inclusions, 198

Stress—contd.
    concentration factor, 222
    concentrations around circular holes in filament stiffened sheet, 189–95
    correlation, 161
    function, 145
    distributions, 221
    field, 132–4, 145, 198
    polarisation field, 185–7
Stresses in
    inclusions, 214
    three-dimensional composites with limiting shear properties, 197–207
Structure factor function, 145
Subsidiary condition equation, 126
Symmetry properties, 143, 144

Taylor series expansion, 104
Tensorial transformations, 180
Ternary composites, constitutive equations for estimating longitudinal properties of, 89
Ternary property maps, 93
Ternary systems, 93, 107, 113
Transformations of variational problems, 126
Transport properties, 171
Transversely isotropic averaged tensor, 182
Trial functions, 122, 130, 131, 144, 159
Tsai–Halpin equation, 160

Unbounded domain, 140
Unidirectional element, role of, 86

Variational formulation for heterogeneous systems, 124–6
Variational methods, 124–32
Variational principles
    application, 154
    based on unspecified reference system, 132–7
    heterogeneous systems, for, 172

Variational problems, transformations
of, 126
Variational treatments, fundamental
theorem for, 144
Vibrations of antisymmetrically
laminated cylindrical shells,
37–54
Voigt
average, 132, 156, 157, 159, 162,
184
model, 160, 162, 168
notation, 182

Volume fraction variables, 92, 106
von-Karman type shear-deformation
plate theory, 64

Weight fraction variables, 92

ξ factor, 160

Young's modulus, 198, 202, 204, 205

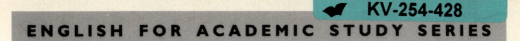
READING

Student's Book

Don McGovern, Margaret Matthews
and S. E. Mackay

Published by
Prentice Hall Europe ELT
Campus 400, Spring Way
Maylands Avenue, Hemel Hempstead,
Hertfordshire, HP2 7EZ
A division of
Simon & Schuster International Group

First published 1994 by Prentice Hall International

© International Book Distributors Ltd, 1994

Typeset in 11/12 Garamond
by Fakenham Photosetting Limited

Printed and bound in Great Britain by
Redwood Books, Trowbridge, Wiltshire

Library of Congress Cataloging in Publication Data

McGovern, Don, 1949–1993.
    Reading : student's book / Don McGovern, Margaret Matthews, and
S.E. MacKay.
        p.   cm. – (English for academic study)
    ISBN 0-13-017872-1
    1. English language – Textbooks for foreign speakers.   2. Reading
(Higher education) – Problems, exercises, etc.   3. Study, Method of –
Problems, exercises, etc.   4. College readers.   I. Matthews,
Margaret.   II. Mackay, Susan E.   III. Title.   IV. Series.
PE1128.M364   1994
428.6'4 – dc20   94-19762
                                                                CIP

British Library Cataloguing in Publication Data

A catalogue record for this book is available from the British Library

ISBN 0–13–017872–1

---

4   5   98

CONTENTS

Introducing this book . . . : 1

Unit 1: Academic success 11

Unit 2: Counselling overseas students 20

Unit 3: Urban development 29

Unit 4: Global warming 41

Unit 5: Education in Asia 54

Unit 6: International diplomacy 66

Unit 7: Development and cultural values in Africa 78

DEDICATION

This book is dedicated to the memory of Don McGovern (1949–1993), writer and poet, whose imagination, care and profundity in wide-ranging concerns of language and the performing arts won him respect wherever he worked.

His work on this book and the Writing volume in this series was completed shortly before his untimely death.

ACKNOWLEDGEMENTS

The authors and publishers would like to thank the following for permission to reproduce the texts in this book:

The Independent for John Emsley 'The hidden good in white bread' (19.3.90).

Routledge & Kegan Paul for Beard and Senior *Motivating Students*, chapter 3 'Students' characteristics and success in higher education' (1980).

The Times Higher Education Supplement for Sian Griffiths 'Talking their language' (1991).

Oxford University Press for Grenyer et al 'The growth of cities' in *Contrasts in Development* (1979).

Blackwell for J. Short 'Cities as if only capital matters' in *The Humane City* (1989).

The New Scientist for 'Balancing the carbon budget' (6.1.90).

The Economist for 'How climate changes' (7.4.90).

Finance and Development vol. 15, no. 1, pp 36–7 for John Simmons 'Can education promote development?' (1978).

The Oxford Review of Education vol. 18, no. 1, pp 17–18 for Wang Gungwu, 'Universities in transition in Asia' (1993).

International Affairs vol. 65, no. 1, Winter, for David D. Newson (1988) 'The new environmental agenda: are governments ready?' (1988/1989).

Finance and Development vol. 28, no. 4, pp 10–13 for Mamadou Dia, 'Development and cultural values in Sub-Saharan Africa' (1991).

INTRODUCING THIS BOOK

Aims

This course approaches the reading of academic texts in English in ways which may be new to you. For this reason it will be helpful to understand the aims and objectives of the course and how the reading texts and exercises relate to these.

During your degree course, you will need to read many texts in your subject area in the form of books, articles from academic journals, reports, etc. You may be asked to do a lot of reading in English, and at first this can seem demanding when you are working in a second or a third language. Some of your reading texts will be more important or relevant for you than others. For this reason they will need to be read with greater care and will probably require more than one reading. However, you may want to read other texts more rapidly or selectively because they are less important for your needs. You will also need to learn how to *select* reading materials to extend your knowledge of your subject. Many of these will be used as a basis for writing academic essays.

This Reading course has been developed in response to the needs of students undertaking academic degree courses in English. The units in this book are designed to help you to do the following:

1. Develop more flexible and efficient reading strategies and skills
2. Develop a fuller comprehension of reading texts
3. Learn to identify relevant reading texts in your subject area more effectively.

Practising the many reading strategies and skills in these units will gradually prepare you for the demands of your academic course in English.

Each unit presents texts on a particular subject. The texts contain important examples of the ways in which academic English is used, and there is a variety of exercises to help you understand these. The subject content and features of language use also relate to the work in corresponding units in the Writing course in this series.

Texts

The texts in this Reading course cover a wide variety of subjects and are of general interest. They are *not* written for specialist readers and do not require a specialist knowledge of any subject. They have been chosen because they cover a wide range of topics and a variety of styles of writing.

Some of the vocabulary in the reading texts may at first seem specialised or slightly technical if you are not familiar with the subject. This impression, however, will be misleading. All of the vocabulary items in these texts are likely to be familiar to educated general readers for whom English is a first language. Terms which are not widely used and may seem slightly technical can be described as *sub-technical vocabulary*. These are forms of vocabulary which are common to many different subject fields and would be found, for example, in quality newspapers and magazines intended for general readers.

Similarly, even though the content or ideas in a particular text may represent a specific subject area, the methods of organisation and features of language use will be found in reading texts in a wide range of academic subjects. In other words, each of the reading texts has been chosen as an example of writing which is in some way typical or representative.

As you progress through the units in this book, you will see how writers present their ideas and how they use language in particular ways to do this. You will then be given the opportunity to practise presenting your own ideas in similar ways in the essays which come at the end of each unit. Each essay topic is designed to relate directly to the reading texts in that unit and gives you the opportunity to select information from these texts to incorporate in your writing. There is also the opportunity in these essays for you to incorporate information from other reading texts, if you wish.

Reading strategies and skills

This course will give you the opportunity to develop and practise reading strategies and skills which can be applied to all forms of academic study. The strategies and skills you will practise are as follows:

1. Predicting
2. Skimming
3. Scanning
4. Detailed reading
5. Guessing unknown words
6. Understanding main ideas
7. Inferring
8. Understanding text organisation
9. Assessing a writer's purpose
10. Evaluating a writer's attitude.

1. PREDICTING

Before you read a text in detail, it is possible to predict what information you may find in it. You will probably have some knowledge of the subject already, and you can use this knowledge to help you *anticipate* what a reading text contains.

After looking at the title, for example, you can ask yourself what you know and do not know about the subject before you read the text. Or you can formulate questions that you would like to have answered by reading the text. These exercises will help you focus more effectively on the ideas in a text when you actually start reading.

To help you predict, you may also use skimming and scanning strategies as described below.

2. SKIMMING

Skimming involves reading quickly through a text to get an overall idea of its contents. Features of the text that can help you include the following:

(a) Title
(b) Sub-title(s)
(c) Details about the author
(d) Abstract
(e) Introductory paragraph
(f) First, second and last sentences of following paragraphs
(g) Concluding paragraph.

A text may not contain all of these features – there may be no abstract, for example, and no sub-titles – but you can usually expect to find at least (a), (e), (f) and (g). Focusing on these will give you an understanding of the overall idea or gist of the text you are reading – in other words, a general understanding as opposed to a detailed reading.

Another term for this kind of reading is *surveying*. Surveying can be described as looking quickly through a book, chapter of a book, article from a journal, etc., to decide whether or not it is suitable for your purpose. To decide whether or not a text is suitable, especially if it is a book, you will also need to focus on the following features in addition to those mentioned above:

(a) Edition and date of publication
(b) Table of contents
(c) Foreword
(d) Introduction
(e) Index.

3. SCANNING

When you scan a text, again you look quickly through it. However, unlike skimming, scanning involves looking for specific words, phrases and items of information as quickly as possible. In other words, scanning involves rapid reading for the *specific* rather than the general; for *particular details* rather than the overall idea.

When you read a text, for example, you may want to find only a percentage figure or the dates of particular historical events instead of the main ideas. Scanning will help you find such information more efficiently.

4. DETAILED READING

A second and third reading of a text will also focus on the secondary ideas and details which support, explain and develop the main ideas. This can be described as a more comprehensive reading. It involves a slower and more careful reading process. At this stage you can also try to guess the meaning of unfamiliar vocabulary (see Section 5 below).

5. GUESSING UNKNOWN WORDS

It is unlikely that you will understand 100 per cent of the vocabulary in a text, especially at a first reading. Use first the *context* and then your own knowledge of the subject to help you guess the meaning of unknown words. At your first reading of a text it is usually best not to stop and consult your dictionary. This will interrupt your process of reading and understanding. Often the meaning of unfamiliar words and phrases becomes clear as you continue to read through the text. The dictionary can be used at a later stage.

In using the context to help you guess unknown vocabulary, you can refer first to the immediate context and then to the wider context in which a word is found. The immediate context is the sentence in which a word is found, and sometimes the sentences immediately before and after this. The wider context can include other sentences and even other paragraphs in a text. Both forms of context can often provide important information which will help you guess the meaning of unfamiliar words.

6. UNDERSTANDING MAIN IDEAS

You will practise recognising the main ideas contained within a text. In the process of skimming you will already have identified some of these main ideas. During a second and third reading you can recognise and understand them more fully. Each paragraph will usually contain one main idea, sometimes referred to as the *paragraph topic*.

The reading materials provide several exercises which help you identify and understand the main ideas in a text. Knowing the key points in a reading text is vital in assessing its importance and relevance for your needs. Understanding the main ideas will also lead you to an understanding of a writer's organisation (see Section 8 below).

7. INFERRING

Sometimes a writer will suggest or express something indirectly in a text. In other words, a writer will *imply* something and leave it to the reader to *infer* or understand what is meant. When writers do this, they rely to some extent on the knowledge of their readers – knowledge of a subject or cultural knowledge, for example. Inferring a writer's meaning is sometimes important in the process of understanding a reading text.

8. UNDERSTANDING TEXT ORGANISATION

Writers structure, or organise, their writing in many different ways. Recognising the way in which a text has been organised will help you understand its meaning more fully. A writer may want, for example, to outline a situation, discuss a problem and propose a solution. This will usually result in a particular pattern of organisation. Or a writer may want to compare and contrast two ideas and will choose one of two basic structures commonly used to compare and contrast.

Another feature related to organisation is a writer's use of time. To give an account of events or describe a process, writers will often use a *chronological order*, in which events are recounted in the order in which they have occurred. Other writers will choose to organise an account of events in different ways, perhaps with repeated contrasts between past and present time.

9. ASSESSING A WRITER'S PURPOSE

Once you understand the organisation of a text, you can then recognise the writer's purpose more clearly. The text organisation a writer selects will partly depend upon his or her particular purpose. A writer may want to *inform* or *persuade*, and he or she will select a structure or pattern of organisation according to this purpose.

A writer may also intend to do both of these things in a written text – to inform as well as persuade. In such cases it is often helpful to try to assess which of these purposes seems to be more important or dominant.

10. EVALUATING A WRITER'S ATTITUDE

Writers are not necessarily neutral or objective when they write, particularly if they are trying to persuade readers to agree with their opinions. It is important that you recognise what an author's attitude is in relation to the ideas or information being presented. This is because such attitudes can influence the ways in which information is presented. You will be looking at ways in which a writer's attitude may be identified. You will also practise evaluating how relatively neutral or biased his or her attitude may be.

Sample reading text

Below you will find a sample reading text in which you will practise some of the skills and strategies outlined above.

TASK 1. SKIMMING AND PREDICTING

1.1 Look quickly *only* at the labelled parts of the article below. Looking at the parts of the article which have been labelled will provide you with important information about the text. For example, you will probably see immediately that this is *not* an advertisement for bread.

1.2 Without reading the text itself, what could you guess about its content and the way it will be presented? Write down as many ideas as you can. It does not matter if all your predictions are correct or not. The important thing is to start thinking about the text before you read it.

Title

Sub-title

First paragraph

The hidden good in white bread

John Emsley on the unexpected fibre in foods that we thought were bad for us

A dish of some supermarket own-brand cornflakes will provide you with half a gram of dietary fibre. However, if you chose Britain's
5 best-selling proprietary brand of cornflakes you will get twice as much fibre. Yet both are made from the same flour; the only difference is in the way they are
10 cooked. The traditional high tem-

perature method used by Kelloggs converts some of the carbohydrate into a form that we cannot digest.

Kelloggs started making corn-
15 flakes in the 1870s. But it was only in 1982 that food chemists at Cambridge discovered this previously unknown form of fibre, called resistant starch. Previously,
20 it had been thought that dietary fibre was mainly cellulose, but when the food chemists analysed the fibre content of bread they discovered that it had more fibre
25 than the flour from which it was made – this fibre is not cellulose but resistant starch.

Now it has been discovered in all sorts of cooked foods such as
30 bread, potatoes, pasta, peas, beans and especially rice. Scientists believe that it is a form of dietary fibre and as such may have a beneficial effect on our health.
35 Its presence in rice could explain why the Japanese suffer much less from diseases of the bowel, such as cancer and diverticulosis.

The average Briton eats 240
40 grams (eight ounces) a day of carbohydrate, essential for producing the energy we need to keep warm and move about. Another form of carbohydrate that we are recom-
45 mended to eat is fibre, which is a general term for indigestible celluloses. Ideally, we should try and eat 30 grams (an ounce) of fibre a day to ensure a regular bowel mo-
50 tion. Constipation is thought to increase the likelihood of cancer and other disorders of the lower gut.

White bread and rice were
55 thought to harbour little fibre compared with their brown counterparts. Now food chemists have discovered some despised foods are not as bad as they have been
60 portrayed. Some contain remarkably high levels of resistant starch fibre. All-Bran for breakfast can provide a third of our daily fibre needs, but so can an evening meal

65 of rice and peas.

Resistant starch explains some of the mysteries of cooked food. It was known in the last century that when we heat certain foods we
70 reduce their energy value. In 1978 Robert Selvendran of the Food Research Institute, Norwich, surprised food chemists by reporting that some foods were easier to
75 digest raw rather than cooked.

In 1982 a team led by Hans Englyst of the Dunn Clinical Nutrition Centre, Cambridge, was able to show that these curious ob-
80 servations were explained by resistant starch which is formed on cooking. As its name implies, this resists the digestive enzymes of the stomach and intestine and en-
85 ters the lower bowel with its energy store intact. There bacteria may attack it in the same way that they attack ordinary fibre and release some of its energy but no-
90 where near the amount that the original starch contained.

Colin Berry, head of nutrition and food safety at the Flour Milling and Banking Research Asso-
95 ciation of Chorleywood, has been researching ways in which the resistant starch contents of foods could be deliberately increased. He has shown that resistant starch
100 depends on three things: the amount of amylose starch in the food, a high cooking temperature, and a high moisture content.

Together these allow the amy-
105 lose polymers to aggregate together into crystallites, which are tightly packed molecular structures held together by hydrogen bonds. The insolubility of these
110 semi-crystalline materials and the close packing of the polymers prevents our digestive enzymes from attacking them. Reheating food will only serve to increase the
115 amount of resistant starch, and may not only taste but be better for you.

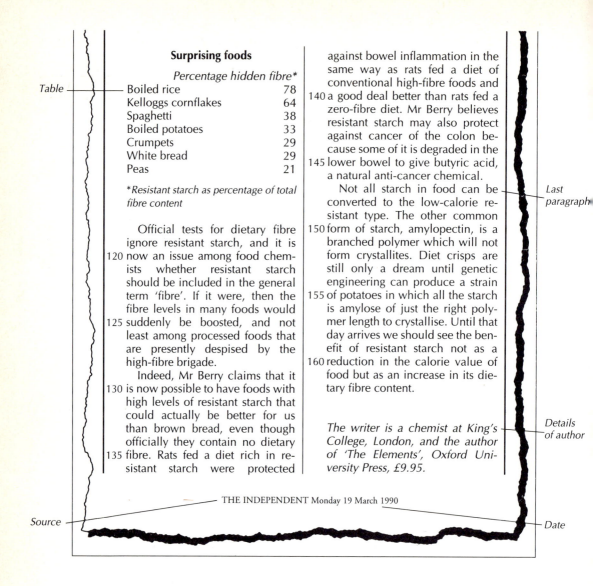

Surprising foods

Percentage hidden fibre*

Table

Boiled rice	78
Kelloggs cornflakes	64
Spaghetti	38
Boiled potatoes	33
Crumpets	29
White bread	29
Peas	21

*Resistant starch as percentage of total fibre content

Official tests for dietary fibre ignore resistant starch, and it is 120 now an issue among food chemists whether resistant starch should be included in the general term 'fibre'. If it were, then the fibre levels in many foods would 125 suddenly be boosted, and not least among processed foods that are presently despised by the high-fibre brigade.

Indeed, Mr Berry claims that it 130 is now possible to have foods with high levels of resistant starch that could actually be better for us than brown bread, even though officially they contain no dietary 135 fibre. Rats fed a diet rich in resistant starch were protected against bowel inflammation in the same way as rats fed a diet of conventional high-fibre foods and 140 a good deal better than rats fed a zero-fibre diet. Mr Berry believes resistant starch may also protect against cancer of the colon because some of it is degraded in the 145 lower bowel to give butyric acid, a natural anti-cancer chemical.

Last paragraph

Not all starch in food can be converted to the low-calorie resistant type. The other common 150 form of starch, amylopectin, is a branched polymer which will not form crystallites. Diet crisps are still only a dream until genetic engineering can produce a strain 155 of potatoes in which all the starch is amylose of just the right polymer length to crystallise. Until that day arrives we should see the benefit of resistant starch not as a 160 reduction in the calorie value of food but as an increase in its dietary fibre content.

Details of author

The writer is a chemist at King's College, London, and the author of 'The Elements', Oxford University Press, £9.95.

Source

THE INDEPENDENT Monday 19 March 1990

Date

Task 2. Understanding main ideas

Unless you are a food scientist, it is likely that there will be many words in the text which you do not know. Some of these you should guess and others you can overlook for the time being. Try to get an impression of the overall meaning and the main ideas, and do not worry at this stage about understanding too many of the details.

2.1 Read the text in order to understand the main ideas. When you have finished, try to complete the summary below:

> There are some foods which scientists used to believe were bad for health, because they did not contain enough _____, (a) for example, _____. (b)
>
> However, recently another substance has been found in them which is very good for health. This is called _____. (c)

You may also be interested in the details of the article, but an overall understanding should come first.

TASK 3. EVALUATING A WRITER'S PURPOSE AND ATTITUDE

To understand a writer's attitude, it is first necessary to ask yourself this question: is the writer's *primary purpose* to inform, to persuade or both?

3.1 Consider the writer and the article:

    (a) What is the writer's primary purpose – to inform or to persuade?
    (b) Do you think the writer probably has a good knowledge of this subject? Why?
    (c) Do you think the writer may have an interest in the sale of white bread?
    (d) Has the article changed or influenced your ideas about food?

TASK 4. A READER'S PURPOSE

Different readers may have different purposes for reading the above article. Below is a list of four people:

    (a) a regular newspaper reader
    (b) a person who buys food for a household
    (c) a food manufacturer
    (d) a scientist.

4.1 What purpose could each of the above people have in reading the article? Consider:

(a) Which reader(s) would have the most practical interest in reading the article?
(b) Which reader(s) would have the most intellectual interest in reading the article?

4.2 How might they use skimming, scanning and detailed reading skills to satisfy those purposes? Consider:

(a) Which reader(s) would be most likely to read the article quickly? Why?
(b) Which reader(s) would be most likely to read the article carefully or give it a second reading? Why?

4.3 In relation to the article, 'The hidden good in white bread', which of the four readers described above would you consider yourself to be most like?

1

ACADEMIC SUCCESS

The topic of this unit involves educational theory. It is based on an analysis of students' attitudes and performance in further and higher education in Britain and the United States.

This unit will give you practice in:

1. Asking predicting questions and answering pre-reading questions about texts.

2. Skimming, scanning and detailed reading.

3. Guessing unknown vocabulary.

4. Understanding the use of generalisation and qualification.

5. Understanding text organisation.

6. Analysing a writer's references to time in text organisation.

7. Understanding the main ideas in a text.

8. Evaluating titles of texts.

9. Identifying a writer's sources.

10. Evaluating a writer's attitude.

11. Understanding the characteristics of a writer's use of language.

PRE-READING TASKS

A. Consider the possible influence of the following on academic success:

(a) Reasons for studying
(b) Physical energy patterns or 'bio-rhythms'
(c) Study methods
(d) Intelligence
(e) Time spent studying
(f) Personality.

Which do you think are the *most* and the *least* important? Why?

B. Until recently *one* of these factors was thought to be very important in academic success. However, a number of studies carried out in Britain have not confirmed this. Which factor do you think it might be? Discuss your choice with another person.

 # Text 1.1

TASK 1

1.1 Now read Text 1.1 quickly and find the answer to question B above.

Text 1.1

Twenty years ago, if teachers in universities and colleges had been asked what individual differences influenced success in higher education, they would almost certainly have mentioned differences in intellectual ability and very probably the effects of interests other
5 than academic ones, such as holding office in the Students' Union, which limited the time that students were willing to give to study. Possibly they would have mentioned study methods. In any case, their answers would have depended on observation and personal belief, since few relevant research results were available in Britain.

10 Today teachers, knowing the growing body of research findings, would be more likely to refer to students' orientations and consequent motivations; to their differing levels of maturity in so far as these influence methods of study and expectations of courses and teachers; to study methods, whether systematic or disorganised;
15 to cognitive styles and personality differences which influence subject choice and may determine how students choose to spend their time; and, in light of some very recent research, they might speculate on possible influences of physiological differences on levels of arousal, application to study or even on the development
20 of one cognitive style rather than another.

Interestingly, informed teachers today would be less likely to mention differences in ability or intelligence in relation to success, for a number of investigations have shown that in the highly selected populations of British sixth forms[1] and colleges, students who do well
25 are usually no more intelligent on average than those who do poorly.

Notes:
1. British sixth forms – British schools or colleges for pupils in the highest section of secondary school, usually entered after the age of 16. Pupils prepare for Advanced Level Examinations ('A' Levels), on the basis of which many go on to higher or further education.

TASK 2

2.1 Read Text 1.1 again more carefully.

2.2 How many different time periods are specified in the text?

2.3 Identify the words which mark or indicate these time periods.

2.4 For each word indicating time, think of a similar word or expression which could replace it in the text.

2.5 How is the text organised? How do the time periods relate to the way the text is organised?

TASK 3

3.1 Below are some words taken from the text. Try to guess their meaning by thinking about the context in which they are found. In each case choose *one* of the three answers which you think best expresses the meaning.

1. orientations   (line 11)
(a) placement or position of something in relation to a map or the points of a compass;
(b) direction of interest towards a particular thing or purpose;
(c) inability to make clear decisions.

2. motivations   (line 12)
(a) reason for doing something; need or purpose;
(b) cause of movement or action;
(c) lack of energy.

3. cognitive styles   (lines 15, 20)
(a) mechanical forms;
(b) different ways of knowing or acquiring knowledge;
(c) emotional expressions.

4. speculate   (line 18)
(a) to buy or deal in goods or shares;
(b) to argue;
(c) to make guesses or form opinions without having complete knowledge.

5. physiological   (line 18)
(a) concerning the functions of the body;
(b) relating to the scientific study of matter and energy;
(c) concerning different mental activities.

6. arousal   (line 19)
(a) a state of being wakened from sleep;
(b) a state of being active or stimulated; alertness;
(c) lack of interest.

7. application   (line 19)
(a) an official request in writing;
(b) the putting of one thing onto another;
(c) careful and concentrated effort or attention.

TASK 4

Text 1.1 contains a number of *generalisations* about the changes in teachers' ideas concerning academic success in higher education.

4.1 Identify at least *three* generalisations in the first paragraph of Text 1.1. The first one has been done for you.

> [Teachers] would ... have mentioned differences in intellectual ability (lines 3–4)

4.2 Which words and phrases are used to qualify or modify the generalisations in Text 1.1 in terms of probability? Consider all three paragraphs in the text. The first one has been done for you.

> **almost certainly** (line 3)

TASK 5

5.1 On the basis of the information given in Text 1.1, is the following conclusion accurate?

> *Intelligence bears no relation to academic performance.*

Why/why not?

5.2 Can you rewrite the conclusion in 5.1 above to make it more accurate?

5.3 What are the limitations of the research in the text? In other words, is it based on information from different kinds of schools in many countries?

Text 1.2

TASK 6

6.1 Consider the following titles:

Cognitive styles
Personality traits
Levels of maturity and students' expectations
Physiological differences and learning

Skim Text 1.2 quickly and decide which one of these titles best expresses the overall idea.

Text 1.2

Teachers at universities sometimes complain that a proportion of students, especially in the first year, show limitations in their thinking. At a number of English-speaking universities, teachers have commented on students' belief that theories should be wholly
5 'right', that a translation should be able to convey the exact meaning of the original, or that experts should never make mistakes. In the event that an exception to a theory was mentioned, or an expert was found wrong in one of his views or explanations, those students became acutely distressed; for when
10 an authority failed them they did not know what to do other than reject it completely. If Bligh is right in his paper on 'The Cynthia Syndrome' (1977), some students arrive in higher education from schools where teachers give high marks for information obtained from books or classroom notes, although no critical faculty is
15 exercised. Indeed, he suggests that some students pass through university – although with a declining standard of performance - never realising that anything over and above acquisition of information is expected of them. When they gain a poor degree, they decide to teach, and so perpetuate this belief.

20   There is some confirmation from early sources concerning limitations in students' thinking. Perry (1968), who investigated the intellectual development of Harvard undergraduates during four years at college, reported three distinguishable stages. In a first stage, he found that students expected to find 'right answers'
25 which were known to 'Authority' and saw it as Authority's role to teach these answers to students. Gibson (1970) found not only that students of sociology had a notion that theories and hypotheses were falsified when they had been shown not to apply in certain circumstances, but they also over-generalised experimental results,
30 failed to recognise particular experiments as building blocks in a

---

wider theory, rejected quantitative data, so regressing large tracts of social psychology back into philosophy, and were confused as to what constituted evidence. He noted a corroborative observation by Peters (1958) concerning students' tendency to seek
35 highly general theories which were logically impossible, and quoted Veness (1968) who spoke of 'concepts so lacking in form and content that they can be conveniently squeezed from the tube in any shape available'.

This seems to correspond roughly with Perry's second stage in
40 which students were willing to accept alternative theories and explanations but tended to swing to the extremes of maintaining that 'everyone has a right to his own opinion'. Alternatively, they might come to the conclusion that there were two or more views which 'Authority' required them to note. Thus, they either
45 assumed that in the absence of an absolute Authority no meaningful judgements could be made, or sought to maintain dependence on Authority and 'absolute truths', since without them they would feel, as one student put it: 'If everything is relative, nothing is true, nothing matters.'

50 By the third stage, Perry found that students realised that knowing and valuing were relative in time and circumstances and that an individual was faced with responsibility for choice and commitment in life. Obviously these differences affect students' learning – those who are still at the first stage will seek right
55 answers to memorise, whilst those at stage three will expect to be required to think – weighing one view against another – and to take responsibility for their own decisions.

TASK 7

7.1 The writer refers to Perry's three stages of intellectual development in students. Read the text again and select *one* sentence which best summarises each stage.

TASK 8

8.1 List the sources referred to in Text 1.2 and indicate which is the main one. One of the sources has been given for you.

Bligh (1977)

TASK 9

9.1 Below are some words taken from the text. Try to guess their meaning by thinking about the context in which they are found. In each case choose *one* of the three answers which you think best expresses the meaning.

  1. **distressed**   (line 9)
  (a) upset because of something alarming or unpleasant;
  (b) unusually happy;
  (c) surprised by a result that had not been predicted.

  2. **faculty**   (line 14)
  (a) a group of similar subject departments in a university;
  (b) a process of gradual development;
  (c) an ability to do something.

  3. **acquisition**   (line 17)
  (a) the process of learning or developing something;
  (b) the process of selecting and organising;
  (c) an inability to remember.

  4. **perpetuate**   (line 19)
  (a) to oppose forcefully;
  (b) to make or allow something to continue;
  (c) to question without any clear purpose.

  5. **regressing**   (line 31)
  (a) developing something to a more advanced stage;
  (b) adopting an overly familiar attitude towards something;
  (c) returning something to an earlier or less developed condition.

  6. **corroborative**   (line 33)
  (a) similar in some respects;
  (b) supporting or strengthening something with information or proof;
  (c) showing major differences.

  7. **relative**   (line 51)
  (a) measured with reference to or in relation to something else;
  (b) difficult to achieve;
  (c) true, right or relevant in all situations.

9.2 Match each of the following words with its synonym in the vocabulary items above:

supportive	reverting	finding	worried	compared to	power	preserve

TASK 10

10.1 Below is a list of attitudes to information and learning. Refer again to Text 1.2 and decide which of the three stages of intellectual development each attitude describes. Write 1, 2 or 3 next to each attitude. The first one has been done for you.

Attitude	Stage of development
(a) There is only one solution to a problem.	1
(b) Experts know everything.	
(c) Theories are flexible.	
(d) Theories are relative.	
(e) Learning means remembering facts.	
(f) There are no absolute truths.	
(g) Independence of thought is a desirable goal.	
(h) Valid experimental results are widely generalisable.	
(i) Theories can be applied universally.	
(j) All individuals can evaluate facts or theories.	

TASK 11

11.1 Does the writer reveal his own attitude in the text? Or does he remain relatively neutral? Choose specific points from the text to support your view.

TASK 12

12.1 How would you describe the writer's use of language in Text 1.2? Is it formal or informal? Personal or impersonal?

12.2 Who are the readers for whom this text was probably written?

12.3 Where would you expect a text like this to be published?

TASK 13

13.1 Do you agree with the ideas expressed in Texts 1.1 and 1.2? Discuss.

13.2 Would these ideas be accepted in your own culture and subject area? Are they too specific to Western cultural and educational conditions? Discuss.

13.3 What are considered to be the most important factors in academic success in your culture and subject area? Discuss.

TASK 14

14.1 Write an essay on the following topic:

> **Academic success in my culture: the most important factors**

What are considered to be the most important factors in academic success in your culture? Do you agree? Have the reading texts in this unit influenced your point of view?

   You may find it helpful to refer to or quote from these texts. You may also want to refer to other reading texts related to this topic.

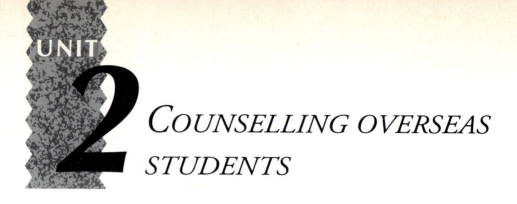

UNIT

2 COUNSELLING OVERSEAS STUDENTS

In this unit you will read two selections from the same article on problems encountered by overseas students in adapting to academic and cultural life in Britain. You will learn how Western psychological counselling has to change to deal with these problems.

This unit will give you practice in:

1. Asking predicting questions about texts.

2. Skimming, scanning and detailed reading.

3. Understanding the main ideas in a text.

4. Applying true/false questions to texts.

5. Guessing unknown vocabulary.

6. Understanding the organisation of texts.

7. Evaluating an individual's degree of certainty in presenting arguments.

8. Understanding the characteristics of a writer's use of language or style.

9. Evaluating titles of texts.

PRE-READING TASKS

Text 2.1 is entitled 'Talking their language'. This phrase is often used idiomatically to mean: communicating with people from other groups or cultures in terms which they can understand; having a common understanding.

A. Consider the title and the description of the topic above. Write down *three* questions that you might expect to be answered by reading this text.

B. Compare your questions with those of a partner and discuss them.

Text 2.1

TASK 1

1.1 Skim Text 2.1 quickly.

1.2 What are some of the main ideas?

Text 2.1

The rain never stopped, the heating failed, the
landlord was racist, I knew no one, my research was
handicapped by linguistic misunderstanding and the
return flight was overbooked.

5   These are the sorts of problems that weekly drive
overseas students into the surgeries of university
counsellors and GPs,[1] with symptoms ranging from
headaches and eye strain to heart palpitations.[2] They
ask for brain scans and eye tests; worry about tumours
10 and cardiac complaints. Counsellors and doctors mark
their files 'stress-related' and attempt counselling.

   But often it is in these surgeries – where help should
be at hand – that the biggest failures occur. For the
psycho-babble[3] of Western therapy, its insistence on
15 finding psychological explanations for physical complaints,
can further bewilder the already alienated student.

   Mr Alex Coren has been practising as a psychotherapist
at King's College, London, for several years. He believes
that: 'Psychological theories that take as their starting point
20 the behaviour of individuals are unlikely to have much to
offer many overseas students. There is a real danger that
we impose our own ethnocentric models on overseas
students – many of whom have belief systems and values
very different from our own.'

25   One problem is that students from certain cultures just do
not believe in the assumptions underlying the therapist's
trade … Likewise, the freedom to pick and choose – and
drop out of – courses is viewed by many government-
sponsored overseas students, whose jobs and families
30 often depend on their academic success, as dangerous and
decadent.

Notes:

1. GPs (line 7) – an abbreviation for General Practitioners, i.e. doctors trained in general medicine without specialising in any one area.

2. palpitations (line 8) – irregular or unusually fast beatings of the heart caused by illness, too much effort, etc.

3. psycho-babble (line 14; informal) – language full of modern slang expressions related to psychology and the awareness of one's own thoughts and feelings.

TASK 2

2.1 Re-read Text 2.1 more carefully. This time concentrate on details as well as main ideas.

2.2 Below is a list of the main ideas in the text, in the wrong order. Re-order the ideas so that they follow the pattern of organisation in the text. The first one has been done for you.

	Paragraph
(a) Beliefs and assumptions not shared by overseas students.	_____
(b) Problems experienced by overseas students.	1
(c) Western counselling imposes Western ideas on overseas students.	_____
(d) Symptoms resulting from these problems and attempted solutions.	_____
(e) The failures of Western counselling.	_____

TASK 3

3.1 Answer **T** if you think the statement is true and **F** if you think the statement is false. Correct any false statements so that they express accurately what is in the text.

(a) The problems experienced by overseas students are caused only by academic study. _____

(b) University counsellors and doctors deal with these problems effectively. _____

(c) Mr Alex Coren has counselled overseas students at Kings College, London, for many years. _____

(d) The terminology of Western counselling always confuses overseas students who are experiencing problems adjusting to British culture. _____

(e) Mr Coren does not agree with most Western psychological theories. _____

(f) Many government-sponsored overseas students regard the freedom to choose and change academic courses as dangerous. _____

TASK 4

4.1 Below are some words taken from the text. Try to guess their meaning by thinking about the context in which they are found. In each case choose *one* of the three answers which you think best expresses the meaning.

    1. stress-related (line 11)
    (a) connected with another department;
    (b) related to the pressure caused by problems with living, working, etc.;
    (c) requiring serious medical treatment.

    2. bewilder (line 16)
    (a) to reassure; to give a sense of security;
    (b) to amuse intensely;
    (c) to confuse, especially by the presence of many different things.

    3. alienated (line 16)
    (a) made to feel isolated or separated from something; made to feel friendless;
    (b) having a secure sense of belonging to something;
    (c) made to know exactly what to expect.

    4. ethnocentric (line 22)
    (a) highly developed;
    (b) experimental and not yet fully accepted;
    (c) believing that one's own race, nation or group is better than others.

    5. decadent (line 31)
    (a) in a state of decline or moral decay;
    (b) specific to only one culture;
    (c) involving high risk but potentially rewarding.

4.2 Match each of the following words with its synonym in the vocabulary items above:

deteriorating	**perplex**	**estranged**	**nationalist**	**psychosomatic**

TASK 5

5.1 In which paragraph does the writer quote or cite someone else's point of view?

5.2 Explain this argument or point of view in your own words.

5.3 For what *two* reasons is the writer worth taking seriously? In other words, why are his/her views likely to be authoritative?

5.4 Find in the text at least *two* examples of beliefs or values held by counsellors which are not usually shared by overseas students.

Task 6

6.1 In lines 17–24, which word does the writer use to introduce the speaker's opinion or point of view to the reader? How much *certainty* is expressed in this word?

6.2 Can you think of words similar to this which could be used in academic writing? How much certainty does each one express? Write them down in order of degree of certainty.

6.3 Which word expresses the degree of *probability* in the speaker's view? How much certainty does this word imply?

6.4 Can you think of words similar to this which could be used in academic writing? Write them down in order of probability. (Refer again to the work you did in Unit 1, Task 4.)

Task 7

7.1 Who seems to be speaking in paragraph 1?

7.2 Why do you think the writer chose to begin the article in this way?

 Text 2.2

Task 8

8.1 Scan the text quickly for the following information:

    (a) To what audience was Mr Coren speaking?
    (b) From what parts of the world were the two students who received counselling from him?

Text 2.2

Mr Coren has counselled students from cultures in which feelings such as fear and loneliness are not viewed as part of an individual's make-up – but are
35 associated with 'supernatural spirits and ghosts'.

'Values I hold as obvious – the right to happiness, autonomy, freedom – are not shared by all overseas students. The world of feelings is linked to spirits, outside presences, not linked to oneself,' he said.
40 Thus solutions designed to maximise the possibility of achieving happiness and autonomy may be quite inappropriate.

His message, to a gathering of university and polytechnic student counsellors, was simple. Before
45 suggesting that the headaches and double vision of the overseas student are something to do with worry about work, family absence or racial harassment, counsellors should first try to understand the students' cultural idiom. 'The societies which send overseas students
50 must be studied and understood if counselling is to be effective,' he said.

Mr Coren told the annual conference of the Association of Student Counsellors that he had reached his conclusions after several cases resisted his
55 psychological approach.

Mr Coren treated an American student on a Rhodes scholarship to London who had previously been seen, at home, by a witch doctor.

Another student, Mohammed F, a 30-year-old post-
60 graduate student from the Middle East, left his post as a university academic as well as his wife and two small children to study for a doctorate in Britain. He was referred to Mr Coren by the university doctor after complaining of double vision and headaches.
65 Mohammed thought he might have a brain tumour and requested a scan. Mr Coren suggested there was a psychological dimension. 'Could the headaches be the result of stress?' Mohammed was startled and upset. A subsequent appointment with Mr Coren was not kept.
70 'The counsellor's role should perhaps be as an intermediary between the old and new cultures,' said Mr Coren.

TASK 9

9.1 Re-read Text 2.2 more carefully.

9.2 What are the main ideas?

TASK 10

10.1 Below are some words taken from the text. Try to guess their meaning by thinking about the context in which they are found. In each case choose *one* of the three answers which you think best expresses the meaning.

1. make-up   (line 34)
(a) powder, colour, etc., worn on the face to improve appearance;
(b) a combination of qualities in a person's character;
(c) one's physical body and its component systems.

2. autonomy   (line 41)
(a) the right to manage one's own affairs;
(b) the state of being free from moral considerations;
(c) the condition of being wholly dependent on others.

3. harassment   (line 47)
(a) a gradual process of integration;
(b) act of causing worry or unhappiness by creating trouble;
(c) severe confusion.

4. idiom   (line 49)
(a) a set of features which are common to many groups;
(b) conflict;
(c) the ways of expression typical of a particular group.

5. intermediary   (lines 70–71)
(a) a person who serves to disguise or conceal the differences between two things;
(b) a person who tries to prevent exchange;
(c) a person who brings two persons, groups or systems into agreement.

10.2 Match each of the following words with its synonym in the vocabulary items above:

dialect	independence	aggravation	personality	go-between

TASK 11

11.1 In Mr Coren's view, how can the failures of Western counselling be overcome (lines 32–51)?

11.2 How did he come to these conclusions?

TASK 12

12.1 Which word emphasises Mr Coren's degree of certainty about the views expressed in lines 32–51? How much certainty is expressed?

12.2 Which words express Mr Coren's degree of certainty in his concluding remarks to the conference? How much certainty is expressed?

TASK 13

13.1 How would you describe the writer's use of language in Texts 2.1 and 2.2? Is it formal or informal? Personal or impersonal? Would you describe these texts as academic writing?

13.2 Who are the readers for whom these texts were probably written?

13.3 Where would you expect texts like these to be published?

TASK 14

14.1 Does the title 'Talking their language' give specific information about the main ideas in this article?

14.2 Write a suitable sub-title to give the reader further information.

TASK 15

15.1 How many of your predicting or pre-reading questions have been answered by reading these texts?

TASK 16

16.1 Do you agree with Mr Coren's conclusions in lines 70–2? Can a counsellor from one culture effectively help someone from another culture? Discuss.

16.2 Is Western psychological counselling practised in your culture, either in academic or non-academic contexts? How effective is it? What are some of the opinions of it in your country? Discuss.

16.3 What are the alternatives to Western counselling in your country? To whom do people go if they need to talk about problems? Discuss.

TASK 17

17.1 Write an essay on the following topic:

> **Can a counsellor from one culture effectively help someone from another culture? Why/why not? What are the factors involved in this question?**

Is a counsellor likely to be effective, for example, if he or she has visited the student's country or studied the student's language before being approached for help? What qualities in the counsellor's personality are helpful in effective counselling?

You may wish to include information from the reading texts in your essay. You may also want to refer to other reading texts which you have found relating to this topic.

UNIT 3

Urban development

In this unit you will read three texts about urban development. The first is taken from a human geography textbook and the second and third are both taken from a book about cities. The first text describes the growth of modern cities throughout the world, while the second and third describe some of the history of urban and industrial development in Britain.

This unit will give you practice in:

1. Answering pre-reading questions about texts.

2. Skimming and detailed reading.

3. Interpreting graphs and tables of information.

4. Understanding the main ideas in a text.

5. Inferring information from a text.

6. Guessing unknown vocabulary.

7. Identifying contrasts in a text.

8. Analysing a writer's references to time in text organisation.

9. Evaluating and choosing titles.

10. Identifying connectives and understanding cohesion in a text.

11. Identifying a writer's expression of attitude.

12. Understanding patterns of organisation in descriptions involving comparison/contrast.

Pre-reading tasks

A. Do you know which three cities in the world are:

(a) the largest in terms of population?
(b) the fastest-growing in terms of population?

B. What problems are associated with large, overcrowded cities?

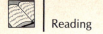

 # Text 3.1

TASK 1

1.1 Skim Text 3.1.

1.2 Then compare your answers to the second pre-reading question with the information given there. Are there any differences?

Text 3.1

THE GROWTH OF CITIES

More and more of the world's population are living in towns or cities, and the speed at which cities are growing in the less developed countries is alarming, as you can see in Figure 3.1. Between 1920 and 1960 big cities in developed countries increased two and a half times in
5 size, but in other parts of the world the growth has been eight times their size.

The sheer size of growth is bad enough, but there are also very disturbing signs of trouble in the comparison of percentages of people living in towns and percentages of people working in industry. During
10 the nineteenth century cities grew as a result of the growth of industry. As you can see from Table 3.1, in Europe the proportion of people living in cities was always smaller than the proportion of the workforce working in factories. Now, however, in the newly industrialised world, the reverse is almost always true: the percentage of people
15 living in cities is much higher than the percentage working in industry.

Without a base of people working in industry, these cities cannot pay for their growth; there is not enough money to build adequate houses for the people that live there, let alone the new arrivals. People arrive
20 in São Paulo, for example, at a rate of 150 an hour. There has been little opportunity to build main drainage or water supplies, or to provide the services such as refuse disposal which a healthy urban life demands. So the figures for the growth of towns and cities represent proportional growth of slums and shanty towns,[1] of unemployment and
25 under-employment, a growth in the number of hopeless and despairing parents and starving children; and yet the cities continue to grow and, undeterred, millions flood into the cities every year.

Notes:
1. shanty towns (line 24) – (parts of) towns made up of roughly built houses of thin metal, wood, etc., where poorer people live.

Figure 3.1

Largest and fastest growing cities in the world

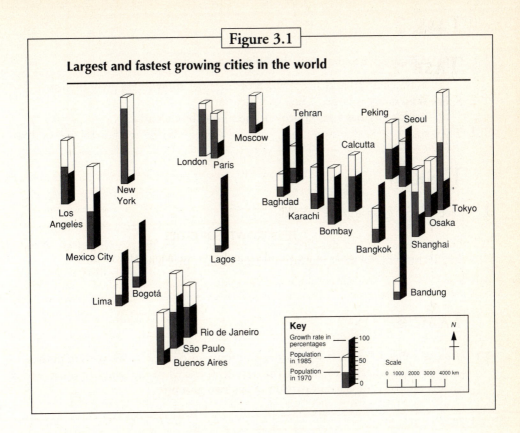

Table 3.1

Percentage of people living in towns and working in industry

Date	Country	Percentage of population living in towns of over 20 000 people	Percentage of workforce in manufacturing industry
1856	France	10.7	22.0
1888	Switzerland	13.0	45.0
1890	Sweden	10.8	22.0
1890	Austria	12.0	30.0
1956	Tunisia	17.5	6.8
1957	Malaya	20.0	6.9
1960	Brazil	28.1	9.5
1961	Venezuela	47.2	8.8

TASK 2

2.1 Look at Figure 3.1 and answer the following questions:

    (a) Between 1970 and 1985 which of the cities shown here grew the least?
    (b) Which city is larger, Lima or Calcutta?
    (c) Which city has a faster growth rate, Lima or Calcutta?

TASK 3

3.1 Read Text 3.1 again more carefully.

3.2 The writers of the text describe various problems in rapidly growing cities. What do they identify as the most important cause of such problems?

TASK 4

4.1 Look at Table 3.1, which compares two groups of countries. What were both groups of countries doing during the periods specified in the table? In other words, what is the basis for the comparison of the two groups?

TASK 5

5.1 Which phrases are used in Text 3.1 to refer the reader to the figure and the table?

5.2 Can you think of similar phrases used to refer to figures, tables and diagrams?

 # Text 3.2 and Text 3.3

These texts are from the same article, entitled 'Cities as if only capital matters'.

TASK 6

6.1 Consider the title of these texts. Do you think the writer believes that money or property are the only things of importance? Why/why not?

6.2 Now read Text 3.2 below. Has your interpretation of the title been confirmed or not?

Text 3.2

In the early part of the eighteenth century, most countries were predominantly agricultural societies. The population was scattered throughout the country in farms, villages and small towns, largely dependent upon human and animal power and the energy of
5 wood, water and wind. Such manufacturing as existed was mainly, though not exclusively, the work of craftsmen, and markets were places where people exchanged or traded goods on a personal basis.

As the cradle of the Industrial Revolution, Britain provided the
10 earliest example of the changes to come. Here capitalist industrialism, with its new machines and new forms of power based on coal and steam, had a marked effect on the places where people lived and worked. From the late eighteenth and throughout the nineteenth centuries increasing numbers of men,
15 women and children were both pushed off the land and drawn away from the farms and country areas into the industrial towns, cities and ports. By the middle of the nineteenth century the rural-urban balance had firmly swung in favour of the urban centres. The countryside may have been bleak for the labouring
20 classes but the cities were no better. The Industrial Revolution created cities dominated by private greed often at the expense of broader social values, community concerns and human dignity. The following extract from Engels's *The Conditions of the Working Class in England*, written in 1884–5, provides only one
25 illustration of the conditions:

In the lower lodging-houses, ten, twelve, sometimes twenty persons of both sexes, all ages and various degrees of nakedness, sleep indiscriminately huddled together upon the floor. These dwellings are usually so damp, filthy, and
30 ruinous, that no one could wish to keep his horse in one of them.

TASK 7

7.1 Read Text 3.2 again more carefully.

7.2 Complete the following sentences based on the information in the text:

(a) Before the Industrial Revolution _____ was scattered.

(b) After machine power was introduced people began to move from

_____.

(c) By the middle of the nineteenth century in Britain more people lived in

_____.

(d) Conditions in the cities were _____ for labourers.

TASK 8

8.1 Below are some words taken from the text. Try to guess their meaning by thinking about the context in which they are found. In each case choose *one* of the three answers which you think best expresses the meaning.

1. predominantly (line 2)
(a) to a limited extent;
(b) completely or entirely;
(c) mostly or mainly.

2. cradle (line 9)
(a) the place where something began;
(b) the place where something was firmly rejected;
(c) a delayed reaction to a new development.

3. bleak (line 19)
(a) pleasant and inviting;
(b) not hopeful or encouraging; depressing;
(c) offering a limited opportunity.

4. greed (line 21)
(a) a desire to do things for the community;
(b) a strong and usually selfish desire for money, power, etc.;
(c) a strong imagination.

5. indiscriminately (line 28)
(a) without careful thought or choice;
(b) cheerfully or happily;
(c) carefully arranged in groups.

6. ruinous (line 30)
(a) located far from the city centre;
(b) having more than enough space;
(c) causing or likely to cause destruction.

8.2 Match each of the following words with its synonym in the vocabulary items above:

randomly	dismal	birthplace	disastrous	largely	avarice

TASK 9

9.1 The paragraphs in Text 3.2 often describe *contrasts*. The contrasting items sometimes appear in the same sentence and sometimes in different parts of the text.
Look at the table below. For each item on the left, suggest another from this text which contrasts with it. The first one has been done for you.

(a) agricultural societies (line 2)	(a) capitalist industrialism (lines 10–11)
(b) new machines (line 11)	(b) _____
(c) the energy of wood, water and wind (lines 5–6)	(c) _____
(d) land ... farms ... country areas (lines 15–16)	(d) _____
(e) urban (line 18)	(e) _____
(f) countryside (line 19)	(f) _____
(g) private greed (line 21)	(g) _____

TASK 10

10.1 How many different time periods are specified in Text 3.2?

10.2 Identify the words which mark or indicate these time periods.

10.3 How is the text organised? How do the time periods relate to the way the text is organised?

TASK 11

11.1 Read Text 3.3 quickly. Then decide which of the following sub-titles best expresses the overall idea in the text.

(a) The continued rise of capitalism
(b) The defeat of capitalism
(c) The control of capitalism

11.2 Which reading skill(s) have you just practised?

Text 3.3

The rise of capitalist industrialism involved the development of a
capitalist ideology, which extolled the virtues of the division of
labour[1] and the efficiency of the 'invisible hand' of perfect competition
35 and self-interest. The market was likened to a finely balanced
mechanism, and the individual worker was seen as a cog in the
machine of production performing a specialised but anonymous
task. In time, specialism, competition and the measurement of
merit by price, became not a description of the way in which markets
40 might behave under 'perfect' conditions but the intellectual foundations
of a 'natural' order of things, a belief, almost a creed, which held that
competitive, selfish behaviour brought about the greatest good for all.

To be sure, the dominant ideas did not go unchallenged. Throughout
the eighteenth and nineteenth centuries different models of society
45 were espoused by social commentators as varied as Tom Paine,
William Morris, Peter Kropotkin and Karl Marx. Their ideas were taken
up by various utopian, anarchist and socialist groups. The alternative
theorists and opposition groups challenged the rule of capital and by
their actions put limits on capitalist forms of development. The
50 industrial cities of the nineteenth century were also places where
labour organisations flourished, self-help schemes such as building
societies were established and communities banded together to resist
the worst excesses of a rampant capitalism. Eventually legislation
was enacted that improved urban conditions, partly through the
55 influence of social critics, partly through the fear of social unrest.
The worst excesses were tempered but the market system continued,
with its commitment to competition, production and growth.

Notes:
1. division of labour (lines 33–4) – a system in which each member of a group specialises in a different type of
work.

TASK 12

12.1 Read Text 3.3 again more carefully.

12.2 Would you still choose the same sub-title for Text 3.3? Why/why not? Refer to
specific points in the text to support your choice.

TASK 13

13.1 Below are some words taken from the text. Try to guess their meaning by thinking about the context in which they are found. In each case choose *one* of the three answers which you think best expresses the meaning.

> **1. extolled**   (line 33)
> (a) highly praised;
> (b) exacted payment from;
> (c) discouraged.
>
> **2. merit**   (line 39)
> (a) economic value;
> (b) the quality of deserving praise or reward; excellence;
> (c) a disagreement among people who work together.
>
> **3. creed**   (line 41)
> (a) a point of view;
> (b) a system of beliefs or principles;
> (c) an illusion which affects more than one person.
>
> **4. espoused**   (line 45)
> (a) explained;
> (b) contradicted;
> (c) supported.
>
> **5. rampant**   (line 53)
> (a) widespread and impossible to control;
> (b) growing;
> (c) declining or deteriorating gradually.

13.2 Match each of the following words with its synonym in the vocabulary items above:

propounded	unrestrained	religion	worth	glorified

TASK 14

14.1 Identify words and phrases in Text 3.3 which contribute to the *cohesion* of sentences – that is, which contribute to the smooth 'flow' of ideas by linking parts of sentences. The first two have been done for you.

> **which** (line 33)
> **and** (line 36)

14.2 Which of these indicate a progression in time?

14.3 Which of these are also markers or indicators of contrast?

Task 15

15.1 Does the writer reveal a specific attitude towards capitalism in Text 3.3? Or is he relatively neutral in his treatment of the subject?

Task 16

A description based on comparison and/or contrast can be developed in two ways. You can group the main ideas on Subject A in one paragraph or section and the main ideas on Subject B in the next paragraph or section, in a 'vertical' movement as shown below:

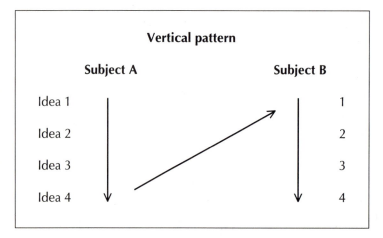

You can also treat the corresponding ideas on Subject A and Subject B as a pair and compare or contrast them one after the other, in a 'horizontal' movement as shown below:

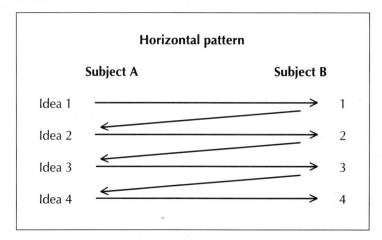

Whether you choose the 'vertical' or the 'horizontal' pattern depends on the kind of text you are writing, its purpose and your own preference. Some writers and readers find the 'horizontal' pattern clearer because it repeatedly reminds them of the comparison or contrast relationship. Others prefer the 'vertical' pattern because of its relative simplicity. The 'horizontal' pattern is often more suitable for a longer piece of writing. In any event, both patterns are common in descriptions involving comparison/contrast.

16.1 Does the paragraph below follow a 'vertical' or 'horizontal' pattern of organisation?

> Apples are generally oval in shape. They range in colour from green to yellow to red. Their texture is usually firm and sometimes even hard. Most oranges, however, are more round in shape. Their colour range is more limited – from vermillion to pale orange. In terms of texture they are relatively soft.

16.2 Rewrite the paragraph in 16.1 so that it follows the opposite pattern of comparison/contrast.

16.3 Which pattern of organisation do Texts 3.2 and 3.3 follow? Treat them as separate texts in giving your answer. Give specific reasons for your choices.

TASK 17

17.1 Which institutions, organisations or places in your home city or capital city serve the interests of people rather than the interests of capital? Discuss.

17.2 What are the most serious problems associated with overcrowding or population growth in the largest city in your country? Discuss.

TASK 18

18.1 Write an essay on *one* of the following topics:

> **Compare and contrast one institution or organisation which serves the interests of people and one which serves the interests of capital in your country.**

OR

> **Compare and contrast the problems associated with population growth in the largest city in your country with those of a major city in another country.**

You may want to refer to the reading texts in this unit. You may also want to include information from other texts on urban development.

Refer again to the 'vertical' and 'horizontal' patterns of comparison and contrast in Task 16 before you do this essay, and arrange your points according to either of these patterns.

UNIT

4 *GLOBAL WARMING*

In this unit you will read three articles about global warming – the warming of the earth by means of the greenhouse effect. They are all taken from journals, and Texts 4.2 and 4.3 are from the same article.

This unit will give you practice in:

1. Asking predicting questions and answering pre-reading questions about texts.

2. Skimming, scanning and detailed reading.

3. Interpreting titles of texts.

4. Guessing unknown vocabulary.

5. Recognising formal definitions and writing extended definitions.

6. Evaluating the degree of certainty and precision in texts presenting measurements.

7. Evaluating a writer's purpose: to inform or persuade.

8. Understanding and re-ordering the main ideas in a text.

9. Understanding the characteristics of a writer's use of language.

10. Evaluating a writer's conclusions.

11. Understanding the **Situation→Problem→Solution→Evaluation** pattern of organisation.

PRE-READING TASKS

A. What do you already know about global warming? Write down notes on what you can remember about this.

B. Discuss your notes with a partner.

C. What would you like to learn about global warming? Write down at least *one* question about global warming which you would like to have answered.

1. assimilate (line 7)
(a) to avoid something carefully;
(b) to understand something completely;
(c) to take in and make use of something.

2. turnover (line 10)
(a) a movement in, through and out of something;
(b) a slow process of reduction;
(c) a danger which is understood by only a few people.

3. trickle (line 18)
(a) a sudden and unexpected appearance;
(b) a thin flow or movement which is usually slow;
(c) a development which increases the expectation of success.

4. shaky (line 22)
(a) not safe or reliable; uncertain;
(b) precise and demonstrated by research;
(c) imaginative and interesting.

5. nutrients (lines 27, 29)
(a) water temperatures measured near the surface;
(b) chemicals or substances providing what is needed for life and growth;
(c) things which cause illness or disease.

6. secretions (line 29)
(a) substances which can be used as drugs or medicines;
(b) things which are taken in slowly and gradually;
(c) materials produced or given off by part of a plant or animal, usually in liquid form.

4.2 Match each of the following words with its synonym in the vocabulary items above:

foods	unsteady	dribble	discharges	absorb	cycle

TASK 5

Definitions can be written in many ways. The most common forms of *formal definition* contain three elements:

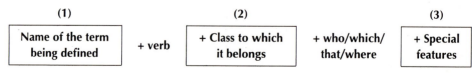

(1)		(2)		(3)
Name of the term being defined	**+ verb**	**+ Class to which it belongs**	**+ who/which/ that/where**	**+ Special features**

Example:

      (1)   (2)   (3)

A triangle is a geometric figure which has three straight sides and three angles.

5.1 Identify the formal definition in Text 4.1. What are its three elements?

Task 6

6.1 The writer of Text 4.1 presents several measurements in the form of numbers. Have these measurements been proved or confirmed by scientific research?

6.2 Look through the text and find any words or phrases which are used to show that the numerical figures are used *tentatively* or *approximately*. The first one has been done for you.

      About 563 gigatonnes (lines 2–3)

6.3 What do these words and phrases show about the nature of scientific knowledge on this topic?

Task 7

7.1 Do you think the main purpose of the writer of Text 4.1 is:

      (a) to provide readers with information, or
      (b) to persuade readers to agree with an opinion?

Discuss.

 # Text 4.2 and Text 4.3

Task 8

8.1 Skim Texts 4.2 and 4.3 quickly to get an impression of the overall idea and the writer's point of view.

8.2 Which of these does the writer propose?

      (a) Radical action
      (b) Limited action
      (c) No action

Text 4.2

HOW CLIMATE CHANGES

Only two things are certain about the world's climate – and one of
them is that it will always be unpredictable. The forces that govern it
are preposterously intertwined, linking the chemistry of the deep
oceans to the physics of the stratosphere,[1] the ice-fields of the Arctic to
5 the forest canopies of the tropics. It is as difficult to find a single cause
for a climate change as it is to pinpoint the sneeze on which to blame
an epidemic of influenza. Such complexity leads to the other
certainty about climate: it is in perpetual flux.

The change that concentrates the minds of politicians and scientists
10 today lies in an abundance of carbon dioxide. This colourless,
tasteless and non-toxic gas, constantly produced by every animal
around, seems an unlikely source of worry. But it has special talents:
it is a much better trapper of heat than are the gases that make up the
bulk of the atmosphere. that much is a certainty of molecular physics.
15 There is no doubt that carbon dioxide – as well as several other gases,
including methane and the man-made CFCs[2] – enhances the
atmosphere's greenhouse effect. Because its level has been
rising ever since people took to burning wood, coal and what-have-you
in industrial quantities, the amount of heat being trapped by it must be
20 increasing.

This greenhouse effect[3] is not controversial. Neither is it synonymous
with global warming. Evidence from the ice ages shows that the
carbon-dioxide greenhouse has always played a crucial role in shaping
the climate, but only as one part of a symphony of effects which few
25 people pay attention to … Other factors may redistribute extra heat in
the atmosphere so as to leave temperatures at ground level
unchanged. Spreading high wispy clouds might reflect sunlight back
into space. Other mechanisms may need study, too, as climatic
theories evolve. New discoveries always bring in their wake modish
30 theories to explain them.

The politicians who will gather in Washington in the coming week for
President Bush's meeting on climate change[4], will hear the
climatologists' current estimates: a warming of between two and six
degrees Celsius before 2100. They will also hear how utterly
35 uncertain those estimates are. The uncertainties will not go away:
political decisions about global warming will always be made on a
teetering balance of probabilities. If and when the warming starts
(it is not clear whether any greenhouse-caused warming has taken place
so far) its future rate and effect will remain in doubt. Arguments about
40 whether greenhouse warming is already happening miss the point. It
does not matter whether this May's darling buds[5] are shaken by
greenhouse winds or ordinary ones. What matters is whether future
droughts will kill the trees they bloom on.

People who seek to shape the climatic future should bear in mind the
45 past. This reveals that climate change, once triggered, tends to gather
a momentum of its own. Falls in atmospheric carbon dioxide – smaller

than the rises predicted for this century and the next – appear to set off
ice ages. Once carbon dioxide levels are up, they stay up for centuries.
Another message from the past is that even if a 'greenhouse world' in
50 which many places are warmer and damper may be attractive, getting
there will not be. Most environments do not take kindly to sudden
changes. Forests take time to migrate, deserts do not bloom overnight.
Even a couple of degrees Celsius of warming in a century would still be
much quicker than anything since the end of the most recent ice age.
55 Rich countries may be able to ride out such ecological upheavals.
Poor ones may not.

Such lessons from the past need to be mixed with tomorrow's
uncertainties: it may never happen. For now, the wisest course is to let
the risk of global warming act as an extra spur for doing only those
60 things that would be wise even if there were no such risk. These
include banning the ozone-chomping CFCs, encouraging cleaner and
more efficient engines, saving tropical forests from the torch and saw,
exploring energy sources like the sun and the wind, and improving
nuclear power. To go beyond such prudence is a large step,
65 betokening the beginning of a new endeavour: planet management.

Notes:
1. stratosphere (line 4) – the outer part of the air which surrounds the earth, between about 10 and 60 kilometres above its surface.
2. CFCs (line 16) – chlorofluorocarbons, chemicals which are used as coolants (e.g. for refrigerators and air conditioners).
3. the greenhouse effect (line 21) – the gradual slight warming of the air surrounding the earth because heat cannot escape through its upper levels; this is thought to be intensified by increasing amounts of carbon dioxide and other gases.
4. President Bush's meeting (line 32) – an intergovernmental meeting of 18 countries on climatic change, held in April 1990. It was a US government initiative and not organised by the Intergovernmental Panel on Climatic Change (IPCC), an international body set up in the late 1980s to look at the same issues.
5. this May's darling buds (line 41) – an ironic reference to 'the darling buds of May' in Shakespeare's Sonnet 18.

Text 4.3

THE ONE-MILLENIUM MANAGER

Man has often changed nature – wiping out a species here, stripping a
region of woodlands there – but in a piecemeal sort of way. Now people
wonder whether to wield the power that two centuries of technology
have given them and have a shot at global air conditioning. Waging
70 war on carbon dioxide by cutting down on the use of fossil fuels would
be only a first step. The climate changes constantly, and once man
thinks he has seized the controls to avert some disastrous man-made
alteration, he will be forever jiggling them to steer his world through
natural changes that will follow.

75   If today's politicians worry about the greenhouse effect, their
     successors the day after tomorrow will have to worry about ice ages.
     The next ice age is probably nearer in time than the death of Julius
     Caesar, perhaps closer than the coronation of Charlemagne.[1] Such
     stretches of time sound thankfully huge to politicians. But if they
80   start to play at planet management they will need to learn to think in
     centuries.

Notes:

1. nearer in time than the death of Julius Caesar ... Charlemagne (lines 77–8) – in other words, between about 2000 and 1000 years from now.

TASK 9

9.1 Read Text 4.2 and Text 4.3 again more carefully.

9.2 Below is a list of the writer's main points, in an order which is different from that in Text 4.2 and Text 4.3. The points represent the main ideas or topics in each of the eight paragraphs in the two texts.

Rearrange this list of points so that it follows the order of main ideas in Text 4.2 and Text 4.3. The first one has been done for you.

		Paragraph
(a)	Action to control climate should be limited in view of uncertainties.	_____
(b)	There are too many natural factors for humans to manage the environment on a large scale.	_____
(c)	Predictions about the effects of the warming of the earth will be very uncertain.	_____
(d)	The world's climate is always changing.	1
(e)	The level of heat trapped by carbon dioxide is rising.	_____
(f)	Any grand plan to control the environment will necessitate a long time scale.	_____
(g)	Climatic changes in the past have had considerable consequences.	_____
(h)	Carbon dioxide is only one of a number of causes of global warming.	_____

TASK 10

10.1 Below are some words taken from Text 4.2 and Text 4.3. Try to guess their meaning by thinking about the context in which they are found. In each case choose *one* of the three answers which you think best expresses the meaning.

1. intertwined   (line 3)
(a) in opposition to each other;
(b) joined or twisted together;
(c) not always consistent or compatible.

2. flux   (line 8)
(a) a state of continuous or constant change;
(b) a condition of being fixed and stable;
(c) a process of growth which is not entirely understood.

3. modish   (line 29)
(a) complex;
(b) overly simplistic;
(c) fashionable.

4. momentum   (line 46)
(a) a quality that is only gradually revealed;
(b) a slowing down of growth;
(c) a force that increases the rate of development of a process.

5. upheavals   (line 55)
(a) great changes, especially with confusion and sometimes violence;
(b) long periods of stable and unchanging conditions;
(c) minor problems which require widespread co-operation.

6. spur   (line 59)
(a) a consideration that would cause one to hesitate;
(b) an event or influence that encourages action;
(c) a long-term benefit.

7. prudence   (line 64)
(a) bold and decisive action;
(b) carefulness in thinking before taking action;
(c) management which is based on the concept of flexibility.

8. avert   (line 72)
(a) to encourage something indirectly;
(b) to report in detail on a well-known event;
(c) to prevent something happening.

10.2 Match each of the following words with its synonym in the vocabulary items above:

caution	stylish	incentive	interwoven	avoid	disturbances	flow	impetus

TASK 11

11.1 How would you describe the writer's use of language in Text 4.2 and Text 4.3? Is it formal or informal? Personal or impersonal? Refer to specific features of the text to support your answer.

11.2 Can you find any informal words or phrases in Text 4.2 and Text 4.3?

TASK 12

12.1 What contrasts are presented by the author's proposed solutions? Are they consistent with each other?

12.2 In what way does the final paragraph of Text 4.3 refer back to the introductory paragraph of Text 4.2? Why does the author do this?

TASK 13

The following pattern of organisation is often found in texts which present problems and explore what can be done about them:

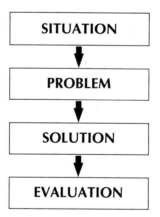

This is a very common way of organising ideas in academic writing. It can be used for:

   a paragraph
   a complete paper or article
   a section of a longer document
   a complete book.

Here is a simple example of this pattern in a short paragraph:

> I am a foreign student living in London. I find it very difficult to meet British students. I will join some university clubs to meet more people. This should help to put me in contact with British students.

13.1 Does the organisation of Text 4.2 and Text 4.3 follow this pattern of **Situation→ Problem→Solution→Evaluation**? Or are there important differences in the way they are organised? Discuss the order and grouping of main ideas in the eight paragraphs in the texts to support your answer. Refer back to the work you did in Task 9 to help you.

TASK 14

In Task 5 you saw one of the commonly used structures of *formal definition* and identified an example of this in Text 4.1:

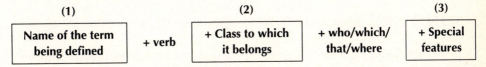

A more extended definition can be written in many ways. Here is one example:

> (1)   (2)
> Revolution may be defined as an event or process in which great, usually
>
>   (3)
> sudden social and political change takes place. This can involve the changing
>
>   (3a)
> of a ruler and/or a political system by force.

The amount and nature of the information you give in a definition will depend on:

whether the concept is new for your readers
how much knowledge of the concept you think your readers will have
how much knowledge of the concept your readers need to have.

14.1 Write an extended definition of global warming using the following information. The items of information have been given in the wrong order.

(a) carbon dioxide, CFCs, methane and other gases
(b) process
(c) global warming
(d) trap heat in the earth's atmosphere
(e) increasing amounts
(f) cause a gradual rise in the earth's temperature

Think carefully about the three basic elements of your definition (refer to the structure at the beginning of this task). Find words and phrases to connect these elements. Then write your extended definition in one or two sentences, using all of the information listed above.

14.2 Compare your extended definition with another person's. Are there important differences?

Task 15

Turn to the pre-reading tasks at the beginning of this unit.

15.1 Has the question you wrote down been answered? If so, what was the answer?

15.2 On the basis of the reading you have done in this unit, write down another question about global warming which you would like to have answered.

15.3 Where can you find further information on global warming to answer this question?

Task 16

16.1 Which environmental issues are important in your country? Do these include global warming? Discuss.

16.2 How important should environmental issues be in relation to the need for economic development? Discuss.

16.3 Many environmentalists claim that the developed world is largely responsible for our current environmental problems. To what extent should developing countries be concerned with global warming and other environmental issues? Discuss.

Task 17

17.1 Write an essay on *one* of the following topics:

> **How important are environmental issues in relation to the need for economic development?**

OR

> **To what extent should developing countries be concerned with global warming and other environmental issues?**

You may want to refer to the reading texts in this unit. You may also want to include information from other texts on environmental issues, especially global warming.

You may want to use the **Situation→Problem→Solution→Evaluation** pattern of organisation in your essay. If so, refer again to Task 13.

UNIT 5

EDUCATION IN ASIA

In this unit you will read two articles about education in Asia. The first discusses the different types of education which can contribute to development. The second surveys the recent history and problems of universities in different Asian countries.

This unit will give you practice in:

1. Asking predicting questions and answering pre-reading questions about texts.

2. Skimming and detailed reading.

3. Understanding definition and extended definition in a reading text.

4. Understanding the main ideas in a reading text.

5. Making a mind map of the organisation of a text.

6. Guessing unknown vocabulary.

7. Recognising reference items and what they refer to in a reading text.

8. Writing formal definitions and extended definitions.

9. Understanding the functions of connecting and cohesive words in a text.

10. Understanding how writers guard against over-simplifying.

11. Recognising transitional paragraphs in a reading text.

12. Understanding the characteristics of a writer's use of language.

PRE-READING TASKS

Text 5.1 was taken from an article entitled 'Can education promote development?' It contains a study of the conditions of education in Pakistan, which are seen to be typical of conditions in many countries.

A. In what ways can education help the social and economic development of a country?

B. Which groups of people are likely to benefit most from efforts to improve education in a developing country?

Text 5.1

TASK 1

1.1 Skim Text 5.1 as quickly as possible. What are some of the main ideas?

1.2 Are these ideas similar to your answers to the pre-reading questions?

Text 5.1

Education can promote development, but it depends on how
development is defined. If it is seen as mainly economic growth,
which tends to benefit upper-income groups, then schooling has
contributed to it by widening the skills and raising the productivity
5 of future workers. If development is defined as mainly improving
the standard of living of the poorest 40 per cent of the population,
then formal schooling has clearly not done much for them, since
most of them are either illiterates or primary school drop-outs.
Moreover, the data show that investment in education widens the
10 gap between the rich and the poor in most countries. This results
from mechanisms like regressive tax systems,[1] expensive
secondary schooling, and free higher education, all of which benefit
mainly the upper-income families. For this article I will define
development as a movement toward a more humane society in
15 both developing and developed nations. Such development
requires political systems more responsive to the interests of the
poor. It also requires rising real income[2] as well as a more equal
distribution and management of wealth.

Types of education

Before looking at some of the issues, a description of the terms
20 used in this article might be useful. *Formal education* or *schooling*
describes the learning that takes place in schools and trains
students mainly for urban, modern-sector jobs. Learning, however,
also takes place outside school, at home, on the street, and on the
job. This is learning by living or learning by doing and can be
25 called *informal education*. One of the few countries which
recognises the importance of informal education during the first
ten years of school is the People's Republic of China; it has
captured some of the benefits of informal learning by getting the
students into informal learning situations on the farms and in the
30 factories. All over the world, professional training in some
disciplines – medicine is an example – has long recognised the
importance of informal on-the-job learning.

Nonformal education is organised learning outside the normal
school university curriculum – examples include training
35 agricultural extension agents[3] in short courses and teaching adults

how to read and write in the evening. Thus nonformal education coexists with formal education, but it receives little funding and less prestige. Upsetting the existing balance between the two is a major source of conflict among educational interest groups.

40   Finally, there is adult *education for self-reliance and participation,* which has its roots both in community development and worker participation in management. Although Paulo Freire, Julius Nyerere, Saul Alinsky and Adam Curle have developed the concept recently, Mahatma Gandhi and others preceded them. Mao Tse-
45 Tung, however, gave the approach its most comprehensive elaboration and application. This approach to education helps groups of people learn how to study together and become aware of the political and economic determinants of their poverty. They then learn to organise and mobilise to improve their circumstances. This
50 differs from the often paternalistic community development approach of the past which relied on outside experts. These adult groups learn that with cooperation and organisation they can build roads, manage water distribution, reduce neighbourhood crime, and grow more food. They learn that they can select their own people to
55 be sent for training as paramedics[4] and teachers. Through cooperative saving they reduce their dependence on money lenders. And when these things happen to them, they develop a self-confidence that in turn generates further initiatives.

Julius Nyerere has emphasised that 'people can only develop
60 themselves; they cannot be developed. Adult education is the key because it will help men and women to think for themselves, make their own decisions, and execute those decisions for themselves.' The rural poor have to transform themselves from being acted upon to being actors.

Notes:
1. regressive tax systems (line 11) – systems in which low-income earners pay a higher percentage of their total income in tax than high-income earners (e.g. someone earning £1000 a year pays £50 i.e. 5 per cent in tax while someone earning £10,000 a year pays £100 i.e. 1 per cent).
2. real income (line 17) – income as measured by purchasing power; the amount of goods and services that can be bought with money earned (even if one earns more money, if prices rise even more, one's real income will fall).
3. agricultural extension agents (line 35) – people trained in new agricultural methods who travel to rural communities to teach farmers about new technology and farming practices.
4. paramedics (line 55) – people who help in the care of ill or injured people but are not doctors or nurses (e.g. ambulance drivers).

TASK 2

2.1 Read Text 5.1 again more carefully.

2.2 Explain in your own words the *two* possible definitions of *development* in lines 1–8.

2.3 Why does investment in education often widen the gap between rich and poor in most countries?

2.4 In which sentence does the writer define *development* for the purposes of this text? Explain this definition in your own words.

2.5 In what *two* ways does the writer extend this definition?

2.6 Would you describe the writer's use of extended definition as formal or not? Refer to Unit 4, Tasks 5 and 14, to support your answer.

TASK 3

3.1 How many forms of education are outlined in lines 19–58? Describe each one simply, in your own words.

3.2 Which form of education is *not* illustrated by examples?

3.3 According to one commentator, which form of education can contribute most to a people's independence?

3.4 How does this person use contrasting verb forms to emphasise his views?

TASK 4

4.1 Complete the mind map below on the basis of information in Text 5.1. It is often helpful to arrange the main ideas in a reading text in this way so that you can understand the organisation of the text more clearly.

The different parts of the mind map below are indicated as follows:

    A = overall idea
    B = main ideas
    C = supporting information

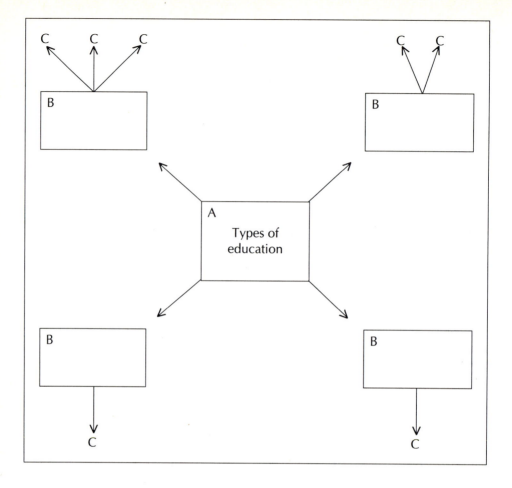

TASK 5

5.1 Below are some words taken from Text 5.1. Try to guess their meaning by thinking about the context in which they are found. In each case choose *one* of the three answers which you think best expresses the meaning.

    1. drop-outs   (line 8)
    (a) unusually successful persons;
    (b) those who leave an institution or withdraw from an activity;
    (c) people who make a limited effort.

    2. humane   (line 14)
    (a) showing kindness, thoughtfulness and sympathy for others;
    (b) concerning people as opposed to animals, plants, etc.;
    (c) well-organised.

3. comprehensive elaboration   (lines 45–46)
(a) systematic opposition;
(b) understanding which is adequate but not fully realised;
(c) complete development with more ideas or details.

4. determinants   (line 48)
(a) conditions which are not directly related;
(b) effects;
(c) factors that settle, limit or decide if something can happen.

5. paternalistic   (line 50)
(a) outdated and no longer relevant;
(b) providing people with what they need but giving them no responsibility or choice;
(c) characteristic of a person who is not strong or forceful enough.

6. initiatives   (line 58)
(a) the first movements or actions which start something;
(b) efforts which can lead to excessive national pride or group loyalty;
(c) gradual economic growth.

5.2 Match each of the following words or phrases with its antonym in the vocabulary items above:

inertia	**democratic**	**results**	**successful people**	**reduction**	**callous**

Task 6

6.1 The following reference items are used in Text 5.1. Each one refers back to something earlier in the text. Explain in each case what the words in bold letters refer to.

(a) **This** results   (line 10)
(b) **it** has captured   (lines 27–8)
(c) between **the two**   (line 38)
(d) **its** roots   (line 41)
(e) **This approach**   (line 46)
(f) **which** relied   (line 51)
(g) **They** learn   (line 54)

Task 7

As you saw in Unit 4, definitions can be written in many ways. The most common forms of formal definition contain the three elements:

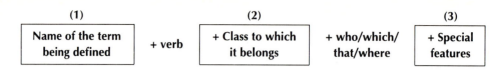

These three elements can also be placed in a different order. This structure is called a *naming definition*:

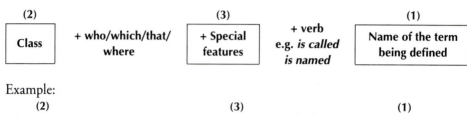

Example:

     (2)   (3)   (1)

The process in which milk is heated to 72°C for 15 seconds is called *pasteurisation*.

7.1 Which of the definitions in lines 19–58 is closest in form to a naming definition?

7.2 Rewrite this as a formal definition following the *first* pattern above.

7.3 Extend this formal definition by rewriting the sentence before it in the text so that it *follows* your new definition. Make any changes that are necessary for clear sense.

 # Text 5.2

Text 5.2 is about the development of Asian universities after World War II.

TASK 8

8.1 Skim Text 5.2 quickly. What are some of the main ideas?

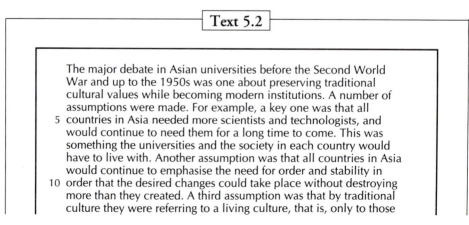

Text 5.2

The major debate in Asian universities before the Second World War and up to the 1950s was one about preserving traditional cultural values while becoming modern institutions. A number of assumptions were made. For example, a key one was that all
5 countries in Asia needed more scientists and technologists, and would continue to need them for a long time to come. This was something the universities and the society in each country would have to live with. Another assumption was that all countries in Asia would continue to emphasise the need for order and stability in
10 order that the desired changes could take place without destroying more than they created. A third assumption was that by traditional culture they were referring to a living culture, that is, only to those

elements of traditional culture which were still meaningful to the lives of the various peoples of Asia.

15   Most university leaders in Asia at the time were aware that the modern university was a product of Western traditional culture and that Western culture itself was changing, having also been modified by the work of the modern universities. Therefore, a close and meaningful relationship existed in the West between the university

20 and traditional culture and there was never really a great gap between the culture which the society wanted to transmit and the values which the university stood for. In the West, it could be argued that even the rate of progress was always regulated by the interaction between the university and the vital sectors of the

25 community. If the society changed too slowly, the universities often led the way; if the university was slow to respond to new social needs, the society sometimes shook it up and prodded it along. The modernisers in Asia admired the university as an institution even though many were aware that the above picture was

30 a rather over-simplified one.

    In Asia, however, there were those who claimed that there had been 'great schools' or a kind of 'traditional university' in the past, but these had been mainly set up to study, enrich and glorify traditional culture (including religious doctrine and practice), or at

35 most (if they were secular) to train officials to preserve the traditional social and political *order*. The modernising elites[1] recognised that this kind of institution was no longer enough. The 'traditional university' would, in modern times, have to be replaced by the modern university in order to help the Asian countries meet

40 the challenges posed by the West during the last hundred years. Thus, by definition, the modern university was seen as a challenge to the traditions which stood in the way of the technological and scientific progress that most Asian governments wanted. Traditional culture was seen by many modernising leaders as

45 opposed to progress, incompatible with science and technology, and therefore something more or less obsolete.

    This is putting the dichotomy in its simplest form and does not do justice to all the people who were debating the issues at the time. But it cannot be denied that many leaders in Asia were

50 divided because of the extreme views they held about the role of universities. There were those who insisted that what was modern was good and desirable and what was traditional was dead and irrelevant; on the other hand, there were those who thought that what was modern was Western, materialistic and subversive (even

55 possibly decadent), and what was traditional was genuine, precious and had to be protected at all costs.

    In reality, the positions taken up by the universities at the time were more varied. Let me sum them up briefly according to the main positions each type of university had taken; these are really

60 ideal positions not necessarily achieved or even achievable under the circumstances. I shall then try and account for these positions by referring briefly to some examples of when and how universities were actually established in Asia.

Notes:

1. elites (line 36) – groups of people considered to be of high importance or status and who have a great deal of power and influence.

TASK 9

9.1 Read Text 5.2 again more carefully.

9.2 Below is a list of the main ideas in Text 5.2, in the wrong order. Re-order them so that they follow the original organisation of ideas in the text. Use numbers (1, 2, 3 etc.) to specify the paragraph you choose in each case.

	Paragraph
(a) The characteristic relationships between Western culture and Western universities.	_____
(b) Divided opinions among Asian leaders on the role of universities.	_____
(c) The main controversy in Asian universities up to the middle of the twentieth century.	_____
(d) The nature of Asian universities and their relationship with traditional Asian culture.	_____
(e) Varied points of view held by those in different types of Asian universities.	_____

TASK 10

10.1 What were *three* of the most important assumptions underlying the debate in Asian universities after the Second World War?

10.2 What did university leaders in Asia think about the relationship between Western traditional culture and its universities? Find *one* key phrase which describes this relationship.

10.3 What were the *two* main functions of the Asian 'traditional university' in the past?

10.4 What did modernising leaders in Asia think about this 'traditional university' of the past? Choose *one* sentence which best expresses this.

TASK 11

11.1 Below are some words taken from Text 5.2. Try to guess their meaning by thinking about the context in which they are found. In each case choose *one* of the three answers which you think best expresses the meaning.

1. **interaction**   (line 24)
(a) cooperation or exchange;
(b) strong conflict or opposition;
(c) a limited effort to engage in a constructive relationship.

2. **secular**   (line 35)
(a) concerned with high academic achievement;
(b) having a clear relation to an existing power structure;
(c) not concerned with spiritual or religious things.

3. **incompatible**   (line 45)
(a) not being a good friend;
(b) not consistent or in agreement with something;
(c) in the best interests of something.

4. **obsolete**   (line 46)
(a) needing further development;
(b) no longer used; out of date;
(c) designed to meet current demands.

5. **dichotomy**   (line 47)
(a) a separation into two parts or things that are opposed;
(b) a situation which is very complex;
(c) a story which will have a wide appeal.

6. **irrelevant**   (line 53)
(a) having qualities which are admired;
(b) not connected with what is happening or being discussed;
(c) challenging.

7. **subversive**   (line 54)
(a) highly desirable;
(b) unusually difficult to understand;
(c) intended to or likely to weaken or destroy something, such as a system or belief.

11.7 Match each of the following words with its synonym in the vocabulary items above:

irreconcilable division interplay unrelated undermining temporal outmoded

TASK 12

The words listed in 12.1 contribute in different ways to the **cohesion** of Text 5.2 – in other words, to the connection and smooth 'flow' of ideas.

12.1 Place the cohesive or connecting words below into *three* different groups, according to the way they work in the text. Use these headings:

1. AND type: Summation Listing	2. BUT type: Concession	3. RELATIVE type

(a) **therefore**   (lines 18, 46)
(b) **while**   (line 3)
(c) **Another (assumption)**   (line 8)
(d) **however**   (line 31)
(e) **Thus**   (line 41)
(f) **A third (assumption)**   (line 11)

(g) **on the other hand**   (line 53)
(h) **In reality**   (line 57)
(i) **For example**   (line 4)
(j) **which**   (lines 21, 22, 42)
(k) **even though**   (line 29)

12.2 Try to describe the function of each of these groups of connecting words in Text 5.2. In other words, try to decide what each group of words has in common.

Task 13

13.1 In three instances in Text 5.2 the writer suggests that the analysis or overview he presents is over-simplified. Which *three* sentences make this clear?

13.2 Can you find a synonym for **in reality** (line 57) in the final paragraph? Why does the writer use such expressions twice in this paragraph?

13.3 Why is the writer careful to explain that these three analytical accounts are over-simplified?

Task 14

14.1 What is the purpose of the final paragraph within the wider context of Text 5.2?

14.2 How would you expect the text to continue?

Task 15

15.1 How would you describe the writers' use of language in Texts 5.1 and 5.2? Is it formal or informal? Personal or impersonal? Describe specific features of each text to support your view.

15.2 For what kinds of readers were these texts probably written?

15.3 Where would you expect texts like these to be published?

TASK 16

16.1 Which of the forms of education outlined in Text 5.1 do you think is the most important? Why? Which one(s) have contributed most to economic and social development in your country? Discuss.

16.2 Is there a university in or near your community? How would you describe its relationship with the surrounding community? How can this relationship be improved? Which contributes more to change, the university or the community? Discuss.

TASK 17

17.1 Write an essay on *one* of the following topics:

> **Which form or forms of education are most likely to contribute to a country's economic and social development?**

OR

> **In your view, what is the ideal relationship between a university and the community around it?**

You may want to begin the topic you choose with an extended definition of important terms, such as 'education', 'development' or 'a university'. The content and structure of your essay on one of these topics will depend on how you choose to understand and define these key terms.

Try to answer these questions on the basis of your experience in your own country or the countries in which you have lived. Specific examples will help to support and illustrate your points.

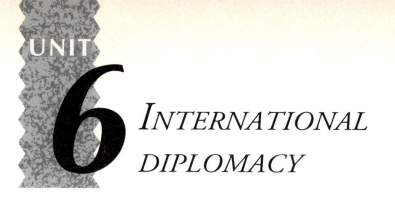

UNIT

6 *INTERNATIONAL DIPLOMACY*

In this unit you will read three extracts from an article about international diplomacy. The article appears in an academic journal about international relations.

This unit will give you practice in:

1. Asking predicting questions about texts.

2. Skimming, scanning and detailed reading.

3. Understanding some of the functions of connecting or cohesive words in a text.

4. Understanding the main ideas in a text.

5. Analysing a writer's references to time in relation to text organisation.

6. Guessing unknown vocabulary.

7. Analysing cause and effect relationships in a text.

8. Identifying the **Situation→Problem→Solution→Evaluation** pattern of organisation in a text.

9. Identifying contrasting ideas in a text.

10. Identifying generalisation and exemplification in a text.

11. Synthesising and summarising information from more than one text.

PRE-READING TASK

The following extracts were taken from an article entitled 'The new diplomatic agenda – are governments ready?'

What do you think might be meant by the phrase, **the new diplomatic agenda?** Discuss your ideas with a partner.

Text 6.1

TASK 1

Text 6.1 is called an *abstract*. You saw this term in the chapter 'Introducing this book'. It is used in academic contexts to describe a short summary of the main points in a text. You will often find abstracts at the beginning of academic articles, especially articles on scientific subjects.

1.1 Skim the abstract below as quickly as possible and see if your prediction about **the new diplomatic agenda** was accurate. Complete this sentence:

'The new diplomatic agenda' in this abstract refers to …

Text 6.1

For most of the twentieth century, the international diplomatic agenda has consisted of questions of political and economic relations between nation-states – the traditional subjects of diplomacy. After the Second World War new diplomatic issues arose, spurred by the technical
5 advances in nuclear energy and electronics. Now in the 1980s, as the century closes, a third set of international problems is emerging: problems that relate to the health of the planet. These new problems will test as never before both the ability of governments to take on new kinds of foreign policy activity and the ability of traditional
10 diplomats to negotiate agreements to meet these challenges. For many of these, the existing nation-state diplomatic relationships will not suffice. New patterns of global management will need to be developed.

TASK 2

The abstract could be seen as falling into three parts: the description of a *situation*, the statement of a *problem*, and the proposal of a *solution*.

2.1 Re-read the abstract. Which sentence states the problem?

2.2 Which sentence proposes the solution?

Text 6.2

TASK 3

Text 6.2 is one of the sections of the article, 'The new diplomatic agenda – are governments ready?' The section is sub-titled, 'National borders become irrelevant'. It deals specifically with the impact of new technology on diplomacy.

3.1 Read the section quickly and tick (✓) to indicate which of the following developments are mentioned. The first one has been done for you.

telegraph	telephone✓	satellite	fax	computer	aircraft
concealed camera		nuclear power	machine translation		television
transport (travel)					

3.2 Which reading skill have you just practised?

Text 6.2

By 1988, therefore, the diplomatic agenda had outgrown the number and type of issues that traditionally occupied Foreign Offices before the Second World War. Not only had the number of nation-states expanded, but diplomacy had become more multilateral; at the same
5   time, the intrusion of diplomats into the internal affairs of other societies had become more common.

It was not just that the tasks of diplomacy had expanded; new technology had accelerated the pace at which nations transacted their business. The communications revolution changed the
10   circumstances in which decisions could be made. In the 1940s and 1950s the telephone and the telegraph made direct and rapid contact between nations easier than ever before. As technology advanced, more and more diplomatic business was conducted by secure telephone – with fewer and fewer records of conversations
15   (future historians will not have an easy time piecing together the events of the latter part of the twentieth century). With the advent of the computer, diplomatic communications increased in speed and in the amount of information carried. Policy-makers were inundated with data as never before; the task of sorting out the important from the
20   unimportant in the mass of reports complicated the processes of assessment and decision.

The modern aircraft meant that national leaders could meet more frequently. The level at which government business was conducted was raised; ministers, not ambassadors, were expected to meet to

25 conduct significant negotiations. The failure of a minister to travel to
    another country to conduct a negotiation was seen as a lack of serious
    intent. The role of ambassadors gradually became limited,
    concentrating on analysis and the delivery of messages, especially in
    the major countries.

30   Television sped not only words, but pictures of events before
    considered diplomatic accounts of the happenings and their
    significance could reach foreign ministries. In the democratic
    countries especially, political pressure often required statements and
    even decisions under the pressure of the public reaction to the
35 television images.

        The greater ease of travel meant greater movement of people and an
    aggravation of older issues. As movement between countries
    increased, governments in both eastern and western hemispheres
    struggled with the issue of minorities. Transcontinental travel
40 accelerated the dissemination of disease: one disease in
    particular, AIDS (Acquired Immune Deficiency Syndrome), became
    an international issue both medically and politically. Those conducting
    political struggles saw in the interruption of diplomacy and
    transportation a means of bringing their cause to public attention.
45 Modern terrorism erupted onto the diplomatic stage, with its hostages,
    hijackings and brutal murders.

        As the 1980s drew to a close, an even more difficult issue arose:
    drugs. The trade and use of narcotics became more widespread,
    partly because travel and communication themselves were so much
50 easier. In both developed and developing countries the spread of
    narcotics affected governments as well as societies. Diplomatic and
    bureaucratic structures throughout the world came under increasing
    pressure to resolve traditional differences over responsibilities and to
    set up interstate machinery that could deal more effectively with the
55 issues involved.

TASK 4

4.1 Which word in the opening sentence of Text 6.2 refers back to something earlier in
the article – in other words, to a previous stage of discussion?

4.2 What else does this word signal to the reader?

TASK 5

5.1 Read Text 6.2 again more carefully.

5.2 Identify the topic or main idea in each paragraph of Text 6.2.

5.3 Identify the time periods specified in the text. Which paragraphs relate to each time period?

Time period	Paragraph(s)	Lines

TASK 6

6.1 Below are some words taken from Text 6.2. Try to guess their meaning by thinking about the context in which they are found. In each case choose *one* of the three answers which you think best expresses the meaning.

1. **multilateral**   (line 4)
(a) extremely complicated;
(b) concerning or including more than two groups or nations;
(c) focused more sharply on fewer issues.

2. **intrusion**   (line 5)
(a) an entrance or interruption which is not wanted or welcome;
(b) an involvement which is welcomed by a majority of people;
(c) a violent disturbance.

3. **accelerated**   (lines 8, 40)
(a) caused or made to move more slowly;
(b) influenced by changes in currency markets;
(c) caused or made to move faster.

4. **intent**   (line 27)
(a) means of transport;
(b) purpose;
(c) sensitivity.

5. **aggravation**   (line 37)
(a) the act of making something worse or more serious;
(b) the return of something to a prominent position;
(c) a means of improvement.

6. **dissemination**   (line 40)
(a) the process of curing or arresting something, especially an illness;
(b) the promotion of something which is attractive or desirable;
(c) the spreading of something widely.

6.2 Match each of the following words with its synonym or its antonym in the vocabulary items above:

improvement	**objective**	**decelerated**	**dispersion**	**unilateral**	**incursion**

TASK 7

7.1 Below is a list of factors that have caused changes in international diplomacy and the ways in which it is conducted. According to Text 6.2, these changes have been caused by new developments in technology. Identify the specific development which has caused each change, in the writer's view.

Factors	**Caused by development of**
Increase in drug use	Transport, computer, telephone
Global terrorism	
Fewer records of conversations	
More ministerial-level meetings	
Increase in speed of communications	
Public pressure for prompt official statements/decisions	
Spread of disease	
Difficulty of assessing importance of new information	

 # Text 6.3

TASK 8

Text 6.3 is a later section of the same article. It is sub-titled 'The new environmental agenda'.

8.1 Skim the text quickly. Describe in your own words what is meant by **the new environmental agenda**.

Text 6.3

Advances in technology have speeded the pace of diplomacy;
post-war international developments have added new issues to the
diplomatic agenda. But their substance is not different in kind from
those that have preoccupied diplomacy for many decades – the issues
5 of peace and war, involving the major powers as well as Third World
nations, economic issues, and the transfer of resources from the richer
to the poorer nations. Even the communications revolution is more of
a change in speed than the introduction of totally different types of
issues into diplomacy. The introduction of the telegraph, which
10 reduced the transit time of diplomatic messages from weeks to a few
hours, may have been even more revolutionary in its time in its impact
on international relations and decision-making than the television and
computer of today. The more public nature of diplomacy today follows
the tradition of statesmen and diplomats of the past who presented
15 their views openly to those of another country. And terrorism aimed
against diplomats is not totally new.

At the mid-point of the twentieth century, the world faced a threat to
its existence in the possibility of nuclear war. That threat has been
contained. The requirement for sophisticated delivery systems for
20 nuclear weapons limited the capacity for mass destruction to a few
nations. Whether the weapons were used depended on the decisions
of a few governments and their leaders. Assuming the worst-case
scenario of a nuclear holocaust is never realised, none of the issues on
the diplomatic agenda up to now has threatened planetary survival.
25 None has threatened mass discomfort and dislocation.

Environmental disaster, unlike the decisions on nuclear weapons, is
caused by the actions of millions of people conducting not war, but
activities they feel are necessary for their health and welfare.

The solutions to environmental issues will involve broad cooperation
30 rather than competition between nation-states, perhaps even
organisations that supersede the nation-state. The problems have
been created by man's exploration of new worlds in space and
biotechnology, and man's use and misuse of an increasingly crowded
planet. The late Barbara Ward, the British development economist,
35 foresaw this in 1971: 'The door of the future is opening onto
a crisis more sudden, more global, more inescapable, more
bewildering than any ever encountered by the human species. And
one which will take decisive shape within the life span of children
already born.' Two American scholars, Harland Cleveland and
40 Lincoln Bloomfield, put the point thus:

Extraordinary realities of modern technology such as rockets,
satellites, jet travel, and computers hitched to tele-
communications, have created a new global agenda. That
agenda comprises, first of all, the old issues in new guises:
45 the management of change without violence, the settlement of
disputes without war. It includes the management of a truly
world economy now close to a nervous breakdown for lack of

institutions that could make the business climate reasonably predictable. In two-thirds of the world, it requires new
50 approaches to development that take account not only of needs for economic growth but of yearnings for a fairer share and passions for cultural identity.

The agenda now highlights also the management of inherently planetary environments (the deep sea, the ocean floor, Antarctica,
55 the weather, outer space), the protection of shared physical resources (soils, forests, fisheries and fresh water), and the global natural cycles such as heat, moisture, energy. Beyond all this, the new international agenda now includes issues that had previously been in the main the province of a country's 'domestic affairs'.

60 The new agenda has been presaged by issues already affecting the health of the planet, such as the question of acid rain between the United States and Canada. The new concerns bring in old problems such as energy, but they also add a new element. The question now is not merely the availability of energy supplies; it is the availability of
65 those energy supplies that will reduce the danger to the health of the planet. Although some precedents exist for treating issues that go beyond the interests of nation-states – such as the UN-sponsored Treaty on the Principles of Outer Space, and the Antarctic Treaty regime preserving the resources of the Antarctic from exploitation –
70 new issues may require new forms of agreement and different knowledge on the part of the diplomat.

Environmental issues have been a matter of concern at least since the 1960s, but they have traditionally been dealt with primarily on a national or regional basis. In 1975, for example, littoral countries
75 around the Mediterranean signed a Convention for the Protection of the Mediterranean against Pollution; the resulting action plan was financed jointly by the signatory countries and the European Community. Public and official attention on a global scale was aroused by scientists' discoveries of ozone depletion, first in the
80 Antarctic and later in the northern hemisphere. This led to a conference in September 1987 of 40 nations in Montreal on the use of fluorocarbons, and the conclusion by 34 of the nations present of a protocol to the 1985 Vienna Convention for the Protection of the Ozone Layer to reduce the emission of chlorofluorocarbons (CFCs) by 35–50
85 per cent by the end of the century.

Concern over the related depletion of tropical forests (at an estimated 30 million acres annually) was manifested in 1986 by the International Tropic Timber Organisation, which has brought together rain-forest nations and the nations that consume the tropical hardwoods.
90 Representatives of governments and private-sector organisations from 42 nations participated in the June 1988 meeting in Rio de Janeiro.

The new global concerns have been further highlighted in 1988 by the attention to the 'greenhouse effect' brought on by record-breaking heat and drought in sections of the United States. Although some scientists
95 have warned of this effect for at least two decades, only the unusual climatic conditions of the 1980s have made it an issue. Typical of the press attention was a headline in *Business Week* on 11 July: 'The earth's alarm bells are ringing'. More formal attention was given at a conference of government officials and scientists from 48 countries,

100 convened by the Canadian government in Toronto on 27 June 1988 on
'The changing atmosphere'. Delegates discussed not only the
greenhouse effect, but also acid rain and ozone depletion. Resolutions
voted called for a 20 per cent reduction in the use of fossil fuels by
the year 2005. Gro Harlem Brundtland, Prime Minister of Norway, called
105 for action. 'For too long … we have been playing lethal games with vital
life-support systems. Time has come to start the process of change.'
On 23 June the US Senate Committee on Energy and Natural Resources
held a hearing on the subject after which two pieces of legislation
were introduced 'aimed at reversing global warming'.

110   The greenhouse effect and acid rain are not the only agenda items
reminding politicians and diplomats of the risks of an overcrowded
world. In this same summer of 1988 residents of the eastern shore of
the United States have been faced with contaminated beaches and
serious problems of waste disposal. Attention has been drawn to
115 depleted rain forests in tropical areas and famine in Africa. The
signs point to a planet that is running out of room for its population.
Political leaders in many countries are now turning to diplomats to
find ways to save the world from the extravagance of its inhabitants;
but diplomats and governments alike may find these problems as
120 difficult to manage and negotiate as any concerning peace or war.
In 1988 it was possible to foresee seven major problems.

TASK 9

9.1 Re-read the first part of Text 6.3 more carefully, from line 1 to line 59.'

9.2 Until recently, what were the *three* most basic issues for international diplomacy?

9.3 Describe *three* ways in which the environmental threat differs from the threat of nuclear war.

9.4 Name *three* specialists to whom the writer refers in support of his argument. What do the views of these specialists have in common?

TASK 10

10.1 Re-read the rest of Text 6.3 more carefully, from line 60 to line 121.'

10.2 In this part of the text the writer outlines the specific issues on **the new environmental agenda**. Write these down in the order in which you find them in the text. The first and last ones have been done for you.

1. availability of less harmful energy

2. _____

3. _____

4. _____

5. _____

6. _____

7. _____

8. _____

9. overpopulation

10.3 Give *one* example of an environmental issue which was dealt with on a regional basis.

10.4 Give *two* examples of global cooperation in dealing with environmental issues.

TASK 11

11.1 Below are some words taken from Text 6.3. Try to guess their meaning by thinking about the context in which they are found. In each case choose *one* of the three answers which you think best expresses the meaning.

1. substance   (line 3)
(a) the real meaning, without the details;
(b) a development which is slow and gradual;
(c) a sudden appearance.

2. supersede   (line 31)
(a) to oppose something without reason;
(b) to enhance the power of something;
(c) to take the place of something used before.

3. presaged   (line 60)
(a) replaced on a gradual basis;
(b) foretold or predicted;
(c) introduced.

4. littoral   (line 74)
(a) near the sea-shore;
(b) warm and semi-tropical;
(c) sharing common economic conditions.

5. depletion (lines 79, 86)
(a) a condition of being increased or intensified;
(b) a state of being unstable;
(c) a condition of being reduced greatly in quantity, size, etc.

6. lethal (line 105)
(a) pleasant and enjoyable;
(b) causing or able to cause death;
(c) containing a small element of danger.

7. extravagance (line 118)
(a) the act of spending or using too much;
(b) the need to feel free and happy;
(c) unusual care in using resources.

11.2 Match each of the following words with its synonym in the vocabulary items above:

essence   foreshadowed   deadly   replace   coastal   running down   wastefulness

Task 12

12.1 Find *three* important contrasts in paragraphs 3–5, lines 26–85. Identify the word or phrase which signals each one. Make a note of the line number.

Task 13

13.1 In which part of Text 6.3 are the main generalisations about **the new environmental agenda** to be found?

13.2 In which part of Text 6.3 are the examples of the new issues on this agenda to be found?

13.3 Look at the final sentence of Text 6.3 (line 121). How would you expect the text to continue?

Task 14

14.1 In a paragraph of 8–10 sentences, summarise the main ideas in Texts 6.2 and 6.3.

TASK 15

15.1 How have new developments in technology and communications affected life in your country in the last ten years? Discuss. (Consider the effects of telephones, computers, fax machines, television, film and video, modern aircraft and rapid transport.)

15.2 Who has benefited most from these changes? Discuss.

TASK 16

16.1 Write an essay on the following topic:

> **The communications revolution in my country: problem or solution?**

In your essay, outline the situation and problems of communications in your country. Do new developments in technology provide solutions? What kinds of problems have these new developments helped to solve? Do they or will they create further problems? How many people can afford these new developments in technology?

UNIT 7

DEVELOPMENT AND CULTURAL VALUES IN AFRICA

In this unit you will read two extracts from the same article about problems with Western development programmes carried out in Africa. The article is taken from a journal about international development and finance.

This unit will give you practice in:

1. Asking predicting questions and answering pre-reading questions about texts.

2. Skimming and detailed reading.

3. Understanding a writer's purpose – to inform or persuade.

4. Identifying a writer's thesis in an essay of argument.

5. Identifying the use of authoritative evidence to support arguments.

6. Identifying contrasting ideas in a text.

7. Guessing unknown vocabulary.

8. Recognising reference items and what they refer to in a text.

9. Understanding the characteristics of a writer's use of language.

10. Understanding the main ideas in a text.

11. Recognising the functions of connecting or cohesive words in a text.

12. Evaluating the degree of certainty in a writer's arguments.

13. Understanding the techniques of introduction and conclusion in a text.

PRE-READING TASK

Both Texts 7.1 and 7.2 were taken from an article entitled, 'Development and cultural values in Sub-Saharan Africa'. The article discusses cultural problems with Western development programmes that have been carried out in this part of Africa.

A. What kinds of problems would you expect to read about in these texts?

B. In what ways can local 'cultural values' be important when countries receive development aid from the West?

Text 7.1

TASK 1

1.1 Skim Text 7.1 as quickly as possible. What are some of the main ideas?

1.2 Do these ideas correspond with your answers to either A or B above?

Text 7.1

Perhaps one of the single biggest hindrances to economic development in Sub-Saharan Africa is the poor performance of the public sector[1] and chronic weaknesses in the local institutions. But decades of efforts by national governments
5 to turn this situation around, with help from the World Bank and other donors, have met with limited success. Recently, however, some observers have come to believe that the problem might well stem from a failure of the traditional approaches to fully integrate the political and sociocultural[2] values that influence
10 economic decision-making. This premise runs counter to the main body of development literature, which tends to dismiss culture as either a neutral element or an obstacle to institutional and technical innovations.

This essay – which emerges from the author's own work, as well
15 as findings of published and unpublished studies and research on the economic psychology of certain ethnic groups in Sub-Saharan Africa – argues that traditional development projects have erred by focusing unduly on technical prescriptions, ignoring the need to adapt development assistance to the local cultural
20 environment and ensure that the Africans identify with such assistance efforts. The thinking behind it forms the basis of a study recently begun by the World Bank, aimed at better integrating traditional cultural traits and incentives in the design and management of projects and programs. The study will (1) explore,
25 through case studies, how sociocultural traits and incentives were linked to the success or failure of various projects and institutions, both in the informal and formal sectors, and (2) use the findings associated with success stories to better manage projects and reform programs dealing with institutional development and
30 management. At this stage, we do not profess to have a full-scale solution to the problem, but we would like to suggest an approach, as part of the current debate on how best to help Africa build efficient and sustainable institutions and managerial capacity.

Notes:
1. public sector (line 3) – the part of a country's economy which is controlled or supported financially by the government.
2. sociocultural (line 9) – involving a combination of social and cultural factors.

79

TASK 2

2.1 Read Text 7.1 again more carefully.

2.2 Is the writer's primary purpose to inform or to persuade?

2.3 Where in the text does the writer state the thesis of his article? Which verb signals this to the reader?

2.4 Explain this thesis in your own words.

2.5 In what *three* ways does the writer suggest that there is a reliable or authoritative support for this thesis?

TASK 3

3.1 In the first paragraph, lines 1–13, the writer presents an important contrast. Describe this in your own words.

3.2 Which word marks or signals this contrast?

3.3 How are different time periods used to emphasise this contrast?

TASK 4

4.1 Below are some words taken from Text 7.1. Try to guess their meaning by thinking about the context in which they are found. In each case choose *one* of the three answers which you think best expresses this meaning.

    1. donors   (line 6)
    (a) persons or groups opposed to development programmes;
    (b) accountants;
    (c) persons who give or contribute something.

    2. premise   (line 10)
    (a) an impression;
    (b) a statement or idea on which reasoning is based;
    (c) a strong and often uncontrollable emotional influence.

    3. erred   (line 18)
    (a) been successful;
    (b) been overly confident;
    (c) made a mistake.

4. prescriptions   (line 18)
(a) statements of what should be done or happen in certain conditions;
(b) orders for medicines written by doctors;
(c) standards of behaviour which have been established over many years.

5. incentives   (line 23)
(a) characteristics which are likely to interfere with growth;
(b) inherited customs or practices;
(c) things which encourage people to greater activity or effort.

6. sustainable   (line 33)
(a) likely to last only a short time;
(b) capable of being kept in existence over a long period;
(c) consistent with local business practices.

4.2 Match each of the following words with its synonym in the vocabulary items above:

do wrong	direction	inducement	benefactor	lasting	assumption

TASK 5

5.1 The following reference items are used in Text 7.1. Each one refers back to something earlier in the text. Explain what each one refers to.

(a) **this situation**   (line 5)
(b) **the problem**   (line 7)
(c) **This premise**   (line 10)
(d) **which**   (line 11)
(e) **it**   (line 21)
(f) **The study**   (line 24)

TASK 6

6.1 Where in the text does the writer refer to himself? What words does he use to do this? Can you think of similar forms of expression?

6.2 How would you describe the writer's use of language or style in Text 7.1? Is it formal or informal? Personal or impersonal? Give specific features of Text 7.1 to support your views.

6.3 In these two introductory paragraphs (lines 1–33) the writer prepares the reader to

expect a particular type of essay in which he will present certain kinds of ideas. What type of essay does he introduce here?

 Text 7.2

TASK 7

7.1 Skim Text 7.2 rapidly. What are some of the most important ideas?

| Text 7.2 |

WHY CULTURE MATTERS

Traditionally, institutional development projects have been based
35 on three key assumptions: a mechanistic and linear conception of
history[1] and 'development', which assumes that every society must
go through the same stages before it can achieve development; a
technological approach to institutional development and
management that assumes that Western methods and techniques of
40 management are the only road to modernisation; and an ethnocentric approach to culture that assumes that the basic goal of any
society is to achieve the same values characterising the so-called
'developed' countries (i.e., spirit of enterprise, profit motive,
material security, and self-interest) – countries not exhibiting such
45 values are viewed as primitive and underdeveloped.

The logical conclusion of this approach is that Africa's
development must be stimulated from the outside, requiring a
transfer of culture, methods, and techniques from the industralised
Western countries. But the evidence to date strongly suggests
50 that none of these assumptions are valid.

First, the remarkable vibrancy of the informal sector[2] in Africa
in the midst of generalised economic crises and difficulties of most
modern sector enterprises illustrates the limits of the linear
conception. Against the background of a hostile environment and
55 lack of government support, the success of most of these micro-
enterprises is best explained by their ability to reconcile African
social and cultural values and traditions with the need for economic
efficiency. Their management is largely a family affair, relying
heavily on informal business relationships.

60 Second, the lack of success of most traditional approaches to
institutional and public sector development in Africa clearly shows
the limitations of the technological approach. In fact, it is now
generally accepted that simply transferring some kind of know-how
(e.g., new teaching methods, new techniques of personnel
65 administration, and new budget devices) to Africa will not suffice.

There must also be a high degree of national identification with the programs and projects – what is now referred to as 'internalisation' or 'ownership'. For this to be optimal, commitments will have to be forthcoming, not just from the political leadership but from the civil
70 servants and general population as well.

Third, extensive studies and research, along with observations by the author, seem to suggest that Western values are not always congruent with traditional incentives and behavioural patterns prevalent in most African countries. Self-reliance and self-interest
75 tend to take a back seat to ethnicity and group loyalty – there are still thousands of ethnic groups on the continent. The main concern seems to be maintaining social balance and equity within the groups, rather than individual economic achievements. Generally, the interest of the local and ethnic communities takes
80 precedence over whatever the government may declare as national goals. Clearly, the six to eight decades of colonisation were simply not long enough for both individuals and governments to develop a new national entity that could transcend ethnicity and the traditional decision-making system.

85   Thus, the need to understand – and take into account – idiosyncrasies of African political and sociocultural structure is of paramount importance if the development community is to help African reform and increase the efficiency of its public and private sectors in a sustainable manner. It is the combination of both the
90 sociocultural and technical–managerial elements of institutional development that will determine the quality and success of efforts in this area.

Notes:
1. a mechanistic and linear conception of history (lines 35–36) – an understanding of history as a coherent 'line' of development (as opposed to a repeated cycle) which can be measured in terms of technological and scientific progress.
2. informal sector (line 51) – the part of a country's private economy which is loosely or informally organised and difficult to register centrally or regulate.

TASK 8

8.1 Read Text 7.2 again more carefully.

8.2 Explain the *three* most important assumptions of traditional development projects. Use your own words as much as possible.

8.3 In the writer's view, what is the result of these assumptions?

8.4 Does the writer agree with these three assumptions of traditional development projects?

TASK 9

9.1 Below are some words taken from Text 7.2. Try to guess their meaning by thinking about the context in which they are found. In each case choose *one* of the three answers which you think best expresses the meaning.

1. vibrancy (line 51)
(a) an inability to function effectively;
(b) a condition in which something can easily be damaged;
(c) a strong quality of energy and enthusiasm.

2. optimal (line 68)
(a) likely to offer at least partial success;
(b) best or most favourable;
(c) incapable of being put into practice.

3. congruent (line 73)
(a) in conflict or opposition;
(b) exercising little or no influence;
(c) similar or consistent.

4. ethnicity (lines 75, 83)
(a) a condition in which the interests of a racial, tribal or national group are given priority;
(b) a lack of secure sense of identity;
(c) a system of ideas placing strong emphasis on individual liberty.

5. transcend (line 83)
(a) to avoid things which seem difficult;
(b) to go beyond or above something, especially normal limits;
(c) to reinforce things which are considered to be usual or conventional.

6. idiosyncrasies (line 86)
(a) habits or ways of behaving which are particular to a person or group;
(b) customs shared by many different groups;
(c) inconsistencies.

9.2 Match each of the following words with its synonym in the vocabulary items above:

surpass	tribalism	peculiarity	power	ideal	corresponding

TASK 10

10.1 How many *counter-arguments* does the writer present to the three key assumptions of traditional development projects? Explain each one in your own words.

10.2 Which connecting words are used to mark or signal the introduction of each counter-argument?

Task 11

11.1 What conclusion does the writer express on the basis of the counter-arguments he presents?

11.2 Which word signals this to the reader?

Task 12

12.1 Write an outline or plan of the main idea or topic in each paragraph of Text 7.2.

Task 13

Writers of academic texts often need to indicate their sources of information and the reliability of that information. They will need to distinguish between supporting information which is accepted or proven and information which is uncertain or has not yet been proven.

13.1 The following list contains phrases which can be used when referring to source material. Refer to the scale in the box below. Using the numbers 1–5, indicate beside each phrase the degree of *acceptance* or *certainty* which it expresses. If you want to locate a phrase between numbers on the scale, you can indicate '2/3', for example. The first one has been done for you:

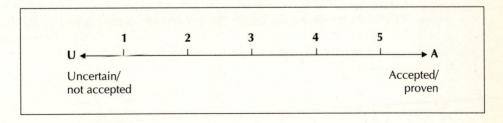

(a) X has postulated that ...   **1/2**

(b) Evidence of this was first reported ...   _____

(c) X and Y have shown that ...   _____

(d) Investigations into ... have revealed little support for ...   _____

(e) A new study by X confirms that ...   _____

(f) X's conclusion is dependent upon the supposition that ...   _____

13.2 In Text 7.1, what sources does the writer give to provide evidence for his ideas?

13.3 The statements below are taken from Text 7.2. They are key expressions of the writer's most important ideas in developing his thesis and counter-arguments.
Refer again to the scale in 13.1. Use the numbers 1–5 to indicate beside each statement the degree of certainty which it expresses.

(a) But the evidence to date strongly suggests that ...
(lines 49–50)   _____

(b) the remarkable vibrancy of the informal sector ...
illustrates ... (lines 51–3)   _____

(c) the lack of success of most traditional approaches ...
clearly shows ... (lines 60–1)   _____

(d) extensive studies and research, along with observations
by the author, seem to suggest that ... (lines 71–2)   _____

13.4 Which of the author's counter-arguments in lines 51–84 is presented with the greatest degree of certainty? Is this degree of certainty consistently maintained in the rest of the paragraph? Find other words and phrases expressing the writer's degree of certainty in the same paragraph.

13.5 Which of these counter-arguments is presented with the least certainty? Is this degree of certainty consistently maintained in the rest of the paragraph? Find other words and phrases expressing the writer's lack of certainty in the same paragraph.

13.6 Which counter-argument is most supported by evidence from research?

TASK 14

14.1 Compare the content of Text 7.2 (lines 34–92) with that of the first paragraph of Text 7.1 (lines 1–13). In what ways are they similar? In what ways are they different?

14.2 Why has the writer organised the text in this way?

Task 15

15.1 Do you agree with the writer's views in these texts? Why/why not? Discuss.

15.2 To what extent should a nation's development be measured in material – i.e. in economic and technological – terms? What other aspects of development would you describe as important? Discuss.

Task 16

16.1 Write an essay on the following topic:

> **To what extent should a nation's development be measured in material terms? What other aspects of development are important?**

In your essay, explore the idea of 'development'. How important are economic growth and technological advances in a country's progress? What disadvantages can these forms of development bring? How important are other aspects of development? What about education? Is a poorer country necessarily 'underdeveloped' if its people have a strong sense of community and a rich cultural life, especially in the arts?

READING

STUDENT'S BOOK

Other titles in the English for Academic Study series:

WHITE, R. and MCGOVERN, D.
Writing

TRZECIAK, J. and MACKAY, S.
Study Skills for Academic Writing

The English for Academic Purposes series:

ST JOHN YATES, C.
Agriculture

VAUGHAN JAMES, C.
Business Studies

WALKER, T.
Computer Studies

ST JOHN YATES, C.
Earth Sciences

ST JOHN YATES, C.
Economics

JOHNSON, D. and JOHNSON, C. M.
General Engineering

JAMES, D. V.
Medicine